The Developmental Ecology
of *Mantispa uhleri* Banks
(Neuroptera: Mantispidae)

The Developmental Ecology
of *Mantispa uhleri* Banks
(Neuroptera: Mantispidae)

KURT E. REDBORG and ELLIS G. MACLEOD

ILLINOIS BIOLOGICAL MONOGRAPHS 53

UNIVERSITY OF ILLINOIS PRESS   Urbana and Chicago

ILLINOIS BIOLOGICAL MONOGRAPHS

Volumes 1 through 24 contained four issues each. Beginning with number 25 (issued in 1957), each publication is numbered consecutively. Standing orders are accepted for forthcoming numbers. The titles listed below are still in print. They may be purchased from the University of Illinois Press, 54 East Gregory Drive, Champaign, Illinois 61820. Out-of-print titles in the Illinois Biological Monographs are available from University Microfilms, Inc., 300 North Zeeb Road, Ann Arbor, Michigan 48106.

KOCH, STEPHEN D. (1974): The *Eragrostis-pectinacea-pilosa* Complex in North and Central America (Gramineae: Eragrostoideae). 86 pp. 14 figs. 8 plates. No. 48. $7.95.

KENDEIGH, S. CHARLES (1979): Invertebrate Populations of the Deciduous Forest: Fluctuations and Relations to Weather. 110 pp. Illus. Tables. No. 50. $12.50.

LEVINE, NORMAN D., and VIRGINIA IVENS (1981): The Coccidian Parasites (Protozoa, Apicomplexa) of Carnivores. 205 pp. Illus. Glossary. Index. No. 51. $15.95.

This monograph is a contribution from the Department of Entomology, University of Illinois, originally submitted in partial fulfillment for the first author's degree of Doctor of Philosophy. First submitted to the Illinois Biological Monograph Board for publication June 1979.

Library of Congress Cataloging in Publications Data

Redborg, Kurt E. (Kurt Eric), 1949-
    The developmental ecology of Mantispa uhleri Banks (Neuroptera: Mantispidae)

    (Illinois biological monographs; 53)
    Bibliography: p.
    Includes index.
    1. Mantispa uhleri—Development. 2. Mantispa uhleri—Ecology.
3. Insects—Development. 4. Insects—Ecology. I. MacLeod, Ellis
G. II. Title. III. Series.
QL513.M3R43   1984   595.7'47   84-32
ISBN 0-252-01085-X (alk. paper)

Acknowledgments

We thank Dr. Stanley Friedman and Dr. Joseph A. Beatty for extensive criticism of the original version of this paper, with special thanks to the latter for his invaluable help with the identification of spiders. We also wish to thank the Illinois Biological Monograph Editorial Committee and two anonymous reviewers for subsequent comments on the manuscript. Dr. Jerome S. Rovner provided timely advice regarding the successful rearing of spiders. Evaluation of the data and statistical analyses were completed with the help of Dr. Arthur Ghent and Dr. Richard B. Selander. To Alice Prickett we extend appreciation for her excellent line drawings of the immature stages of *Mantispa uhleri*.

Some of the work reported here was conducted at the University of Illinois Dixon Springs Agricultural Center. We express our thanks to the director, Dr. C. J. Kaiser, and staff of this institution.

Finally, and most important, the first author wishes to thank his wife Annemarie for valuable discussions and moral support, and to express appreciation to Rachel Anderson of the University of Illinois Press for her painstaking preparation of the manuscript. Without their efforts this work could not have been published.

Contents

Figures

1. Review of the Literature

For more than two hundred years naturalists have been fascinated by members of the neuropteran family Mantispidae because of their remarkable resemblance to the Mantidae, brought about largely by the possession of an elongate prothorax and raptorial prothoracic legs. In fact, early systematists often confused the groups and described new species of mantispids as mantids. This phenomenon is now recognized as an excellent example of convergent evolution—these insects are not closely related phylogenetically but have evolved similar adult structures due to similar selective pressures. The similarities are actually rather superficial, and close examination reveals many profound differences between the groups. Most striking is the fact that preying mantids are hemimetabolous with nymphal stages that closely parallel the adult; mantispids are holometabolous with larval stages structurally and ecologically distinct from the adult.

The family Mantispidae consists of two subfamilies: the Mantispinae and the Platymantispinae. The Platymantispinae will concern us little in this report. Suffice it to say that it is the more primitive subfamily and its members have a less mantid, more lacewing-like appearance. The larval stages of all mantispids are predaceous. Larval platymantispines have been naturally associated with a number of different insects, and in the laboratory several species have been reared on a variety of sedentary arthropod foods (MacLeod and Redborg, 1982). In contrast, all known larval feeding associations for the Mantispinae are with spiders.

Our knowledge of the developmental ecology of the Mantispinae begins with the patient work of Friderich Brauer who, in 1852, published an illustration of the egg and first instar larva of

1

Mantispa styriaca Poda, which had been secured from a rare, field-collected female. Three years later, while excavating in the soil, Brauer discovered a pupa of *M. styriaca* in its cocoon. In retrospect, it seems likely that this cocoon was actually within the egg sac of a burrowing lycosid that had been killed or frightened away by Brauer's digging. It is a testament to his careful methods that his illustration (Brauer, 1855) of the reconstructed appearance of the cocoon in the soil shows it surrounded by what was, unbeknownst to him, the tattered remains of a spider egg sac. It was not until Rogenhofer's (1862) serendipitous observation of the emergence of an adult of *M. styriaca* from the egg sac of a species of *Lycosa* that the pieces of this puzzle began to fall in place for Brauer. Even then, seven more years were to elapse before it became certain that spider eggs were the obligate larval food of this species (Brauer,1869).

Brauer demonstrated that larvae of *Mantispa styriaca* can burrow directly through the wall of a spider egg sac. Seventy years later, when Hungerford (1939) discovered ten to fifteen first instar larvae on the pedicel of a field-collected spider, *Arctosa littoralis* (Hentz), additional information was finally obtained to show how spider eggs can be found by a first instar larval mantispid. Viets (1941) actually observed that first instar larvae of *Mantispa interrupta* Say, released into a container holding an adult female of an unidentified lycosid, boarded it and positioned themselves around the spinnerets. The egg sac produced by this spider yielded a pharate adult of *M. interrupta* and since larvae had ignored egg sacs that had been presented to them, Viets hypothesized that the larvae had entered the sac at the time of its production. Kaston (1938, 1940) also suggested the possibility of a mantispid larva's first boarding a spider, in order to account for his rearing of a specimen of *Mantispa fuscicornis* Banks from an egg sac of *Agelenopsis naevia* (Walckenaer) (cited as *Agelena naevia*) which had been spun after the spider had been collected.

In contrast to this route, McKeown and Mincham's (1948) studies of *Mantispa vittata* Guerin are consistent with Brauer's original observations of the direct penetration of egg sacs. They report that, when confined in a jar with a female spider, larvae made no attempts at boarding. In contrast, living mantispid larvae were discovered within two egg sacs that had been placed in the vicinity of searching first instar larvae, although the penetration of the sac was not observed.

The most recent word (as of 1979; but see Addendum) on the mantispid larval route to its food has come from Lucchese (1955, 1956), who concluded that larvae of *Perlamantispa perla* (Pallas) (cited as *Mantispa perla*, transferred to *Perlamantispa* by Handschin, 1960) were unable to penetrate the egg sacs of lycosids which he offered. The preferred larval targets were the immature spiderlings and adults of lycosids which the larvae boarded and remained upon (although Fig. 60 of the 1956 paper shows not only lycosids but also gnaphosid spiderlings carrying larvae) as the latter matured and from which they entered the egg sacs produced by mature females.

Quite obviously, our understanding of the path by which larval mantispids locate their food has been fragmentary since Brauer's time. We believe that one of the important results of our study is a clarification of this situation.

In addition to the studies of Brauer, Viets, McKeown and Mincham, and Lucchese, two other investigations involved successful rearing of mantispines from the egg. Bisset and Moran (1967) reared an unidentified South African mantispid and described in detail its cocoon-spinning behavior. Davidson (1969) provided a brief description of the larva of *Mantispa viridis* Walker and reared three adults of this species that were fed a novel diet of crushed cabbage loopers, *Trichoplusia ni* (Hübner).

Most of the remaining literature touching on the developmental ecology of these insects has been the result of casual observations, such as the unexpected emergence of a mantispid from the egg sac of a spider that had been collected for quite different purposes. Poujade (1898) and Main (1931) reported the rearings of *Mantispa styriaca* from the egg sacs of *Drassodes hypocrita* Simon and an unidentified gnaphosid [=drassid], respectively. Kishida (1929— original not seen; cited in Bristowe, 1932) described the association of *Eumantispa harmandi* Navas with a clubionid and a ctenizid. Kaston (1940) recorded the emergence of *Mantispa interrupta* from the egg sac of *Gnaphosa muscorum* (L. Koch). Also reporting on the emergence of adults from egg sacs, Milliron (1940) recorded *Cupiennius sallei* (Keyserling) and Parfin (1958) recorded *Agelenopsis* sp. (probably *A. pennsylvanica* C. L. Koch) as spiders whose egg sacs are utilized by *Mantispa viridis*. Stein (1955) described the emergence of two green mantispids (undoubtedly *M. viridis*) from spider egg sacs collected in New Jersey (Stein, personal communication). Although Stein identifies the sacs as that of lycosids, the illustration reveals that they are probably from a clubionid or gnaphosid. Another green mantispid, *M. viridula* Erickson, was reported by Birabén (1960) from the egg sacs of *Metepeira labyrinthea* (Hentz). *Mantispa decorata* Erickson was reared from the egg sac of *Lycosa poliostoma* (Koch) by Capocasale (1971) and, finally, George and George (1975) reported the emergence of *Climaciella brunnea* (Say) from the egg sac of a lycosid, *Tarantula* sp.

The fortunate collection of an adult mantispid, or the findings of a location in, or time of year during which adult mantispids were comparatively common have prompted several reports. Thus, Smith (1934) observed the adults and obtained eggs and first instar larvae of *Climaciella brunnea* (cited as *Mantispa brunnea*) as well as the eggs and first instar larvae of *Mantispa sayi* Banks. Smith's paper also recorded the emergence of two pharate adults of *M. interrupta* from egg sacs of the salticid *Philaeus militaris* Hentz. The collection of numerous adults of *Mantispa interrupta* during one summer made possible Hungerford's (1936) account of mating in this species. This same year Hoffman (1936) also described the eggs and first instar larval behavior of *Climaciella brunnea* (cited as *Climaciella brunnea* var. *occidentalis* Banks). Batra's (1972) description of the behavior of this species suggested that the "host" might be hymenopterous, inconsistent with the rearing observation of George and George (1975). Kuroko's 1961 report provides notes on the eggs and first instar larvae of two Japanese mantispids.

Lastly, while his findings do not fall into any of the above categories, it is worth mentioning that Valerio (1971) studied the natural occurrence of larval *Mantispa viridis* and of the hymenopteran *Baeus* sp. in the egg sacs of *Achaearanea tepidariorum* (C. L. Koch).

Previous studies on the Mantispidae have been hampered by their reliance on happenstance observation rather than upon data derived in significant amounts from controlled situations. It is the purpose of the present study to sharpen our understanding of these remarkable insects through an intensive examination of one particular species, *Mantispa uhleri* Banks. We hope that this endeavor will prove to be as important to future workers as Brauer's investigations were to us.

2. Laboratory Culture of Mantispids

The following materials and methods, which will be used throughout the monograph, have been successfully employed in rearing from the egg some thousand adults representing not only *Mantispa uhleri* but also *M. fuscicornis* Banks, *M. interrupta* Say, *M. pulchella* (Banks), *M. Sayi* Banks, and *M. viridis* Walker.

Maintenance of adults

Initial laboratory cultures used the offspring of two female *Mantispa uhleri* collected at Ferne Clyffe State Park, Johnson Co., Illinois, 25 August 1972 at ultraviolet light. The cultures have been periodically outcrossed to wild-caught stock from a number of Illinois locales.

The mantispids were kept in small screw-top glass jars (5.9 cm diam x 6.4 cm high) with holes (0.2 cm diam) drilled in the Bakelite top for ventilation. The top and sides of each jar were lined with filter paper. Adults were given a house fly each day and had constant access to water from a small cotton pledget which was kept moist. Eggs were readily laid on the filter paper-lined sides and top. The egg clutches, still on the filter paper, were isolated in 2-dram shell vials as soon as detected, and were incubated in a glass chamber over a saturated water solution of KBr that maintains a relative humidity of 80% (Sheldon and MacLeod, 1971).

Larval rearing-chambers

Prior to egg hatching, artificial rearing chambers—"pseudosacs"—were prepared by excavating cylindrical depressions in a hardened mixture of nine parts plaster of Paris to one part

6

powdered activated carbon, the optimal proportions for use in rearing several species of neuropterans (MacLeod and Spiegler, 1961; MacLeod, personal observation) and other small arthropods (Huber, 1958, and references therein). Water was added to the dry mixture until it became just fluid enough to pour in drops, as a series of small globules. A Petri dish served as a suitable mold and container, and, after hardening, the pseudosacs were drilled slightly deeper than wide, by means of a drill bit ¼-inch in diameter. After complete drying, compressed air was used to blow the pseudosacs free of dust.

For rearings, at least three layers of spider eggs were added to each pseudosac. A fine camel's hair brush was used to place one first instar larva in each pseudosac, and the area around the hole was slightly moistened with distilled water. The pseudosacs were then closed by placing over each a 1-cm square of glass cut from a standard 1-mm thick microscope slide. This was gently pressed until a seal was made. These pseudosacs (Figs. 1, 2 and 3) are a modification of the rearing chambers designed by McKeown and Mincham (1948) and later employed by Bisset and Moran (1967).

After all moisture from the wetting procedure had dissipated, the pseudosacs were placed in the 80% relative humidity chamber. It is necessary to remove free water droplets from the vicinity of the eggs, because at 80% relative humidity these droplets will persist long enough to support the growth of mold which is lethal to mantispid larvae.

Larvae were left undisturbed, except for being visually observed through the glass, until engorgement was well under way and there was no possibility of a larva's escaping by slipping out between the glass and the plaster. The chambers were then opened so that more spider eggs could be added.

After being spun, the cocoons (Fig. 3) were carefully removed from the pseudosac and transferred to individual 7-dram vials lined with filter paper. Vials were stoppered with cotton plugs wrapped with Kimwipes tissue, and returned to the humidity chamber. The tissue wrapping prevents the pre-tarsal claws of the newly eclosed adult from becoming entangled in the cotton fibers. During eclosion, the pharate adult bites its way out of the cocoon and climbs up the vial's filter paper-lined side where it undergoes its final ecdysis.

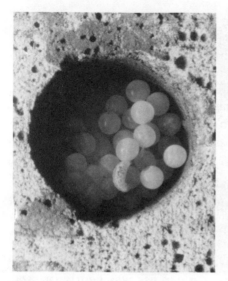

Fig. 1. Second instar mantispid larva in pseudosac.

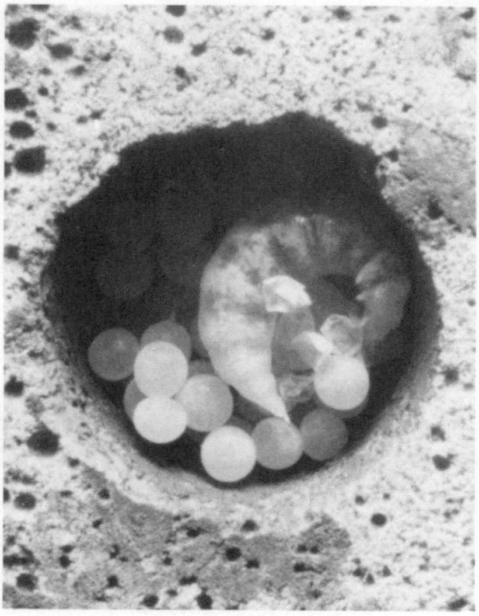

Fig. 2. Third instar mantispid larva in pseudosac.

Spider eggs

Eggs from many species of spiders have proved successful as larval food for *M. uhleri*, and it is likely that most spider eggs would be suitable. For instance, we have used the eggs of *Argiope aurantia* Lucas, which are found in late summer. Because these eggs are generally cemented together within the egg sac, they are suitable as larval food only after being separated from each other, care being taken to prevent any flow of yolk from punctured eggs. Because of their ready availability, however, the eggs most used in our present study were those of a theridiid, *Achaearanea tepidariorum*, an agelenid, *Agelenopsis* sp., and a salticid, *Phidippus audax* (Hentz).

Adult females of *Achaearanea* were collected in numbers during the summer and fall under bridges, under park benches, and around the windows of houses and garages. Females were placed in individual filter paper-lined 7-dram vials stoppered with cotton, and fed house flies, *Musca domestica* Linnaeus, every other day. Egg sacs for immediate use were removed as soon as produced and stored at 5°C to retard development. For long-term use, *Achaearanea* eggs can be killed by freezing and

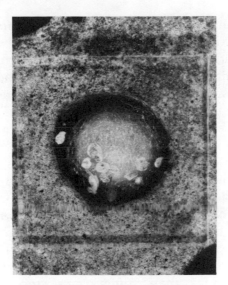

Fig. 3. Cocoon in pseudosac; covering glass slide in place.

stored at -20°C for long periods with no apparent effect on their subsequent suitability as mantispid food.

During the winter months, when *Achaearanea* were not available, egg sacs of *Agelenopsis* sp. were collected from beneath the bark of trees. They were especially abundant beneath the bark of living Osage orange trees, *Maclura pomifera* (Rafinesque) Schneider, which are commonly found in central Illinois as hedgerow plantings. Such eggs, collected in midwinter before development had been initiated, provided a very satisfactory larval food. Unlike *Achaearanea* eggs, however, *Agelenopsis* sp. eggs proved unacceptable as food after they had been killed by freezing.

In the same winter environment where *Agelenopsis* eggs occurred, subadults of *Phidippus audax* were found in silken retreats. These were matured in the laboratory and eggs obtained from females about 2 weeks after mating. These served to span the period from early spring (when *Agelenopsis* eggs were no longer suitable because of their stage of development) to summer (when *Achaearanea* were again present). *Phidippus audux* eggs were also unsatisfactory after freezing.

Care was taken to give first instar mantispid larvae only eggs that showed no signs of development. We have observed that by the time the spiderlings' appendages can be seen through the chorion, the mantispid larvae seem not to be feeding. In contrast, third instar larvae can feed somewhat on even newly hatched spiderlings, although small black marks occasionally appear on the mantispids, suggesting that the spiderlings are capable of damaging the larval integument.

Since Davidson (1969) had some slight success using a non-spider diet of crushed cabbage loopers for *Mantispa viridis,* we attempted to substitute a more readily available larval food for *M. uhleri.* Eggs of the bagworm moth *Thyridopterix ephemerae-formis* (Hayworth) were tried, and, although third instar larvae did feed on them, the moth eggs had a toxic effect. Our experience suggests that an alternative food would probably be no easier to employ than spider eggs, since, with a little planning, an almost unlimited supply of these can be obtained.

Rearing and mating conditions

Egg incubation and larval-pupal rearing took place at a controlled photoperiod of L:D= 16:8, 25°C, and 80% relative humidity. Adult mating and oviposition occurred under these same conditions except that the relative humidity fluctuated between 20 and 50%. Mating behavior was observed under a GE 60 W 115-125 V ruby red light during the early scotophase of the daily cycle. Virgin males and females were used for observation of mating behavior. Pairings were made in plastic cages measuring 8.5 x 12.5 x 6.0 cm with 3.7-cm diameter screened holes cut in the top for ventilation. To reduce cannibalism, the adult mantispids used were well fed and at least 1 week old.

Measurements

Mantispid larvae used for measurements of size and duration of developmental stages were fed eggs of *Achaearanea tepidariorum.* Third instar larvae were presented with an unlimited supply of these eggs so that some were still available when the larvae began spinning cocoons. Second and third instar larvae were measured from material either preserved in Peterson's KAAD fluid and stored in 95% ethyl alcohol or killed in hot water, fixed in Kahle's fluid, and stored in 70% ethyl alcohol. Unfed first instar larvae were macerated in Nesbitt's fluid (Nesbitt, 1945) and temporarily mounted in glycerine beneath a cover slip prior to measurement. Except for the lengths of engorged larvae, all measurements were made under a dissecting microscope with an ocular micrometer, calibrated with a stage micrometer. Because of their curved shape, engorged larvae were drawn on graph paper while being viewed through an ocular grid and were measured with a planimeter.

The drawing of the unfed first instar larva (Fig. 4) was made from a cleared specimen prepared and mounted as described above but examined under a phase-contrast compound microscope. Drawings of second and third instar larvae were made from specimens macerated in Nesbitt's fluid, stained with Chlorazol Black E (2% in 70% ethyl alcohol), and examined in glycerine.

Egg counts were made by enlarging photographs of individual mantispid egg clutches taken subsequent to hatching and after staining the filter paper background with India ink (Fig. 5).

Colors noted in our descriptions were directly compared to a standard color atlas (Maerz and Paul, 1930) and the specific reference in parentheses following each of our subjective descriptions refers to the closest shade in this reference by plate number, column, and row.

All measures of dispersion in this monograph are standard errors of the mean.

Experiment 1: Adult size variation

Since a considerable size variation in field-collected adult mantispids has been observed by numerous workers, an experiment was designed to examine the relationship between the amount of food ingested in the larval stages and adult size attained. Differing allotments of eggs of *A. tepidariorum* were made available to third instar larvae by placing the eggs in the pseudosacs of engorged, quiescent second instar larvae just prior to ecdysis. Allotments of 30, 40, or 50 eggs, or a quantity in excess

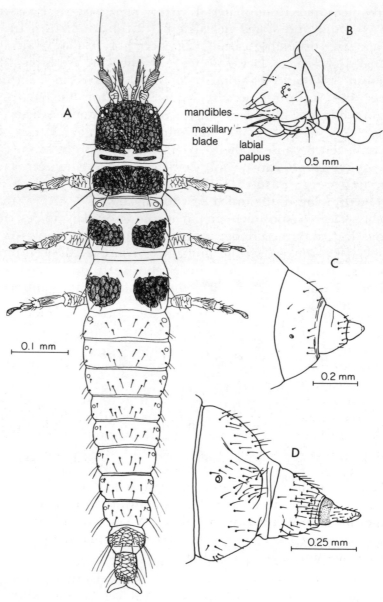

Fig. 4 A. Unfed first instar larva; dorsal view.
   B. Third instar larva; head capsule, lateral view.
   C. Second instar larva; abdominal tip, lateral view, segments
   VIII-X.
   D. Third instar larva; abdominal tip, lateral view, segments
   VIII-X.

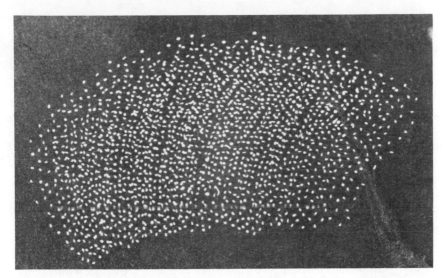

Fig. 5. Egg clutch after hatching.

of what the larva would consume were utilized. For this last group, enough eggs were placed in the pseudosac so that an excess remained at the time of cocoon spinning.

The following criteria served as indices of adult size: the width of the head capsule measured through the eyes at their widest point in anterior view; the length of the forewing from the base of the subcosta to the wing tip; and the length of the pronotum, as seen in lateral view, from the posterior point of articulation with the cervical sclerites to the point of articulation with the mesothorax. These were compared with similar measurements taken from the largest and smallest wild-caught males and females available to us. Head capsule measurements were compared by means of a two-way ANOVA via multiple regression analysis using food groups and sex of the mantispid as main effects, assuming no interaction.

RESULTS

Oviposition

Large individual clutches of eggs were deposited as a series of curved rows produced by the slow back-and-forth movement of the abdomen of the ovipositing female mantispid (Fig. 5). Each egg was attached to the filter-paper jar lining by a short stalk; stalk length varied from clutch to clutch. Ten laboratory-reared females, mated only once, averaged an egg clutch every 3-5 days and produced 12.6 ± 1.7 fertile egg clutches throughout their lifetimes. The mean number of eggs per clutch from four wild-caught females from three different localities ranged from 614 to 2,976 (Table 1). Thus a large female is capable of laying more than 35,000 eggs during her lifetime.

Table 1. Egg production of wild-caught *Mantispa uhleri* females

Female	Eggs/Clutch (Mean ± SE)	% Hatch
1	614 ± 15 $(N = 3)$[a]	98.84 ± 0.45
2	$1,802\pm101$ $(N = 4)$	97.88 ± 0.60
3	$2,307\pm220$ $(N = 5)$	98.37 ± 0.60
4	$2,976\pm149$ $(N = 4)$	98.45 ± 0.37

[a] N = number of clutches.

Egg

Ellipsoid; prior to any embryonic development varying greatly in color from one clutch to another, from tannish yellow (9H2) through pale tan (10B3) to light pink (9A3); eyes become visible at 5.6 ± 0.2 days ($N = 10$ clutches) and the usual neuropteran pattern of transverse banding is visible through the chorion at 7.1 ± 0.2 days ($N = 10$ clutches; each clutch is from a different female). Egg measurements are given in Table 2.

First instar larva (Fig. 4A)

Campodeiform, with a series of transverse reddish brown (6K11) segmented bands on abdomen, each band consisting of a posterior bilateral pair of hook-shaped pigmented areas with the point of the hook directed cephalad and the shank of the hook perpendicular to the dorsal vessel which is visible between the medial ends of the pair; each thoracic segment with a pair of dark brown (7E10) laterodorsal sclerites; background color of the unfed first instar larva light peach (9B5). Average length 0.884 mm. Duration of first stage 7.4 days (Tables 2 and 3).

Table 2.   Measurements of developmental stages of *Mantispa uhleri*

Developmental stage	Measurements in mm (Mean ± SE)
Egg	
Length, micropyle excluded	0.362 ± 0.003(*N*=15)
Width[a]	0.182 ± 0.002(*N*=15)
First instar larva	
Length of unfed larva	0.884 ± 0.008(*N*=7)
Length of engorged mature larva[b]	1.621 ± 0.007(*N*=7)
Width of head capsule[a]	0.126 ± 0.002(*N*=7)
Mature second instar larva	
Length[b]	2.945 ± 0.097(*N*=6)
Width of head capsule[a]	0.243 ± 0.004(*N*=7)
Mature third instar larva	
Length[b]	10.6 ± 0.4 (*N*=7)
Width of head capsule[a]	0.468 ± 0.012(*N*=7)

[a] At widest portion.
[b] Measured from tip of mouthparts to posterior margin of tenth tergum.

Table 3.   Duration of developmental stages of *Mantispa uhleri*

Developmental stage	Duration in days (Mean ± SE)
First instar larva	7.4 ± 0.3(*N*=11)
Second instar larva	2.3 ± 0.1(*N*=11)
Third instar larva prespinning	2.5 ± 0.2(*N*=11)
Third instar larva postspinning	5.7 ± 0.1(*N*=11)
Pupa	9.9 ± 0.2(*N*=11)
Adult[a]	114.0 ± 7.0(*N*=8)

[a] Maximum longevity of mated laboratory-reared females.

Subsequent development of the larva

Prior to ecdysis, the engorged first instar larva passes through an immobile quiescent period during which the cuticle becomes shiny and transparent and the eyes of the pharate second instar larva can be discerned within.

The onset of ecdysis is signaled when a dorsal rent in the old cuticle appears at the rear margin of the head capsule and then proceeds to extend caudad for about one-half the length of the larva. At no time does the old head-capsule split. The method by which the second instar larva frees itself from the old cuticle does not differ appreciably from that described for *Chrysopa oculata* Say by Smith (1922). The newly ecdysed larva is immobile for a short time before it resumes feeding. Ecdysis of the third instar larva follows an identical sequence.

Second (Fig. 1) and third instar (Figs. 2 and 6) larvae

Physogastric, the legs disproportionately shorter and less heavily sclerotized than those of first instar larvae and lacking any important locomotor function.

At a magnification of 30X, only slight morphological differences appear to exist between these instars, the most pronounced involving an increase in the number of setae in the third instar and the modification of the tenth abdominal segment into a spinneret in this form (Figs. 4C and 4D). The color of the second and third instar larvae is determined by the color of the egg contents which they ingest. Larvae fed *Achaearanea* eggs are a pale tan (10B3) with an overlay of white mottlings produced by isolated portions of the larval fat body.

Cocoon (Fig. 3)

Cocoon spinning is basically the same as that described by Bisset and Moran (1967), with the spinneret tracing the same "figure of eight" movements in the final phase of spinning after a "haphazard" foundation has been erected. The internal portion of the cocoon is composed of a series of silken panels, each made up of a number of "figure of eight" figures. Each panel is connected to the one previously made by a strand of silk so that if a cocoon is cut open and one panel withdrawn, a series of panels then follows like scarves from a magician's hat.

Pupa

Exarate, with the prothorax not greatly enlongate; two pairs of dorsal abdominal hooks on each of the third and fourth abdominal segments, similar to those described for the hemerobiid *Wesmealius quadrifasciatus* (Reuter) by Killington (1936) and for

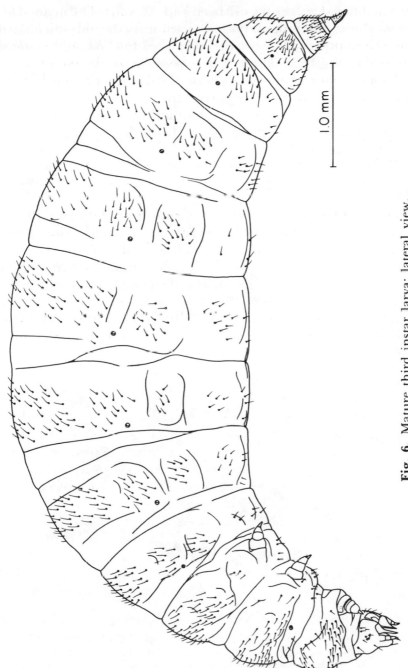

Fig. 6. Mature third instar larva; lateral view.

1.0 mm

18   *Mantispa uhleri* Banks

the mantispid studied by Bisset and Moran (1967); on each segment, the points of the anterior pair of hooks project cephalad and the points of the posterior pair project caudad, the dorsal vessel running between the members of each pair.

After ecdysis to the adult, a dark, liquid meconium is emitted from the anus. This is in contrast to the hard meconial pellet produced at eclosion by other Neuroptera except the Conioptery-gidae (Withycombe, 1925) and the berothid genus *Lomamyia* (MacLeod, unpublished).

Adult (Fig. 7)

Pronotum dark grey (8A9) with a darker mid-dorsal line; femur and tibia of prothoracic leg mostly shiny black (48L7) but, variably, a paler patch beside the femoral teeth laterally and at the proximal end of tibia; coxa pale, with a dark line running the length of its anterior surface; meso- and metathoracic legs pale tan (10B3); abdomen yellow (9J2), marked with black (48L7), a broad black line down the venter, a black line laterally intersecting the spiracles although area directly around spiracles yellow, a black

Fig. 7. *Mantispa uhleri* Banks, adult male.

line down the midline of the dorsum expanded laterally at the posterior margin of each segment.

Sexual dimorphism as follows: Female pterothoracic epimeral plates, episternal plates, mera, trochantins, and coxae black (48L7) with only the area directly around sulci pale yellow (9E2); corresponding parts in the male pale yellow with only occasional splotches of black; black abdominal lines of female somewhat broader, especially the ventral line, so that female abdomen carries proportionately more black than the male.

Mating behavior

The observations summarized here were made from detailed notes taken during six pairings with four successful matings. Subsequent observations of many additional matings have produced no substantial deviations. Two matings were followed closely enough for accurate timing of some of the behavior components (Table 4).

The mating ritual usually begins with the male and female on the top of the cage in an upside down position. Less commonly they are positioned on the side of the cage, or, rarely, on the floor of the cage. Courtship starts with the male and female facing each other. Beginning with either sex, the male and female reciprocally slowly extend forward the coxa and femur of one foreleg, followed by a movement which returns the leg to its original position. This is followed by an identical sequence of movements of the other foreleg. We term these alternating bouts of extension and flexion "sparring." As noted, sparring is usually reciprocal in that the movements of the right foreleg of one sex are followed by identical movements of the right foreleg

Table 4. Timings of epigamic components in *Mantispa uhleri*

Pair number	Abortive sparring cycles	Engagement of genitalia completed		
		Minutes after ♂ and ♀ placed together	Minutes after initiation of sparring	Minutes in copulation
1	0	5.20	0.75	23.25
2	2	9.63	0.72	35.90

of the other, then followed by a similar exchange between the sexes involving the left foreleg, and so on. Sparring will sometimes be stopped and restarted resulting in individual extensions and flexions by either partner, but once both sexes begin doing it at the same time it is always reciprocal. While sparring, the male may repeatedly flick his abdomen up and down. At this time a sweetish odor can be detected. Prior to or between bouts of sparring, male and female may stand closely facing each other, motionless, except for rapid antennal movements by both. Antennal contact does not occur.

After sparring, the male turns away from the female and assumes a characteristic position in front of her in which his abdominal segments are extended, his wings held vertically over the body with their dorsal surfaces together, while both forelegs are simultaneously slowly extended and flexed. We refer to this position as the "wings-up stance." During this time that the male is in front of the female he is usually turned away from her, although his body may be perpendicular to hers with his head to the right or left of her body. At no time is he facing the female, nor does he move while in the wings-up stance. During this time the female continues to spar and may move toward the male.

The male then backs up and, if necessary, rotates his body so as to achieve an orientation parallel to that of the female. He then curves the tip of his abdomen toward the female's genitalia and touches her. If adequate contact is made, the male continues to rotate until his head is pointed in the opposite direction to that of the female. As is usual in the Neuroptera, the tips of the two abdomens have their ventral surfaces in contact so that, when this final orientation is attained, the male's abdomen is twisted 180 degrees. While the male is attempting to make contact, his wings remain with their dorsal surfaces together. After copulation begins, his wings are lowered from their fully raised position, but, like the female's, they are held high enough to prevent interference with copulation.

Upon separation, the male and female turn toward each other and the courtship is often ended with a quick jab of the forelegs, usually from the female. The male can often be seen to apply his mouthparts to his genitalia shortly after mating. A large opalescent spermatophore protrudes from the tip of the female's abdomen after a successful mating. This is no longer visible after 24 hours.

Courtship can be terminated in several ways. The male and female may begin sparring, but one or both may walk away. The female may abandon the male after he assumes the wings-up stance or may back away from him when he attempts copulation. Sometimes several cycles of sparring and of assuming the wings-up stance occur before a successful engagement of the genitalia is achieved.

DISCUSSION

Eggs

The number of eggs laid per clutch is quite variable among female mantispids (Table 1), although the totals vary much less among clutches from the same female. From subjective observations of the egg clutches from many wild-caught females, it appears likely that the number of eggs per clutch and the ovipositing female's body size are directly proportional. Although no measurements were taken, we recall that Female 1 (Table 1) was exceptionally small and by far the smallest of the four females listed.

Size of the female may account for the variable stalk lengths observed, since larger females may lift their abdomens higher from the egg-laying surface before releasing the egg.

Adult size variation

Under culture conditions, third instar larvae begin cocoon-spinning when they have consumed all of the eggs in the pseudosac or when they are unable to locate any of the remaining eggs. After spinning begins, the larvae ignore any additional eggs provided, even though these would have been consumed if given prior to the initiation of spinning. A direct relationship exists between the amount of food ingested by the third instar larva and the body size of the resulting adult (Table 5). This factor in all the estimates of adult size is obvious; therefore, a detailed analysis was performed only for head capsule width.

Larvae fed an unlimited number of spider eggs were nearly twice as large as those fed only 50 eggs and were even larger than any of the others. Because of the disproportionate weight that this class would contribute to the multiple regression analysis, it was excluded from consideration. Among the first three groups, sex$_{adj}$ was not significant ($F_{1,11} = 1.428, 0.5 > P > 0.25$); but food groups$_{adj}$ were very highly significant ($F_{2,11} = 47.204, P > 0.001$, MSE $= .0008$).

This ability of mantispids to reach maturity with an exceptionally wide range of food supplies is in direct contrast with the situation in Chrysopidae and Hemerobiidae (MacLeod, unpublished), in which death occurs to a third instar larva unless it receives a nearly full supply of food.

Only a single third instar *M. uhleri* larva successfully matured on a diet of 30 *Achaearanea* eggs, suggesting that the ingestion of some minimal amount is probably necessary for successful cocoon spinning, pupation, and adult eclosion and that 30

Table 5. Adult size variation in reared and wild-caught *Mantispa uhleri* (Exper. 1)

Eggs provided	Adults measured Number	Sex	Adult measurements (mm) (Mean ± SE) Head capsule width	Wing length	Pronotum length
30	1	♀	1.200	7.5	2.0
40	2	♂	1.300 ± 0.025	7.7 ± 0.0	2.2 ± 0.0
	2	♀	1.312 ± 0.037	8.0 ± 0.3	2.2 ± 0.1
50	3	♂	1.408 ± 0.008	8.1 ± 0.2	2.3 ± 0.0
	7	♀	1.423 ± 0.010	8.9 ± 0.1	2.4 ± 0.0
Excess	3	♂	2.442 ± 0.117	14.3 ± 0.5	4.1 ± 0.3
	2	♀	2.475 ± 0.049	16.2 ± 0.2	4.0 ± 0.0
Representative wild-caught adults[a]	Smallest ♂		1.550	9.0	2.5
	Smallest ♀		1.550	9.9	2.7
	Largest ♂		2.475	15.4	4.1
	Largest ♀		2.375	14.5	4.2

[a] Approximately 100 specimens of each sex were examined.

Achaearanea eggs are close to providing this minimum. Above this minimum, our data suggest that (regardless of the availability of food) the probability of reaching the adult stage is quite high, although the size of the adult produced is closely related to the amount of food consumed.

The ability of a third instar larva to spin after a minimal amount of food has been eaten is assuredly adaptive, for once a larva enters a spider egg sac its food supply is fixed because it has no way of reaching another egg sac. The food content within an egg sac (dependent on egg size and number) is probably variable within any spider species and is most certainly variable among species. Since there is good evidence that under natural conditions *Mantispa uhleri* enters the egg sacs of spiders of a number of species belonging to several families, it is probable that the food supply available for the larvae will be quite variable.

The size variation of these mantispids contrasts with the relatively uniform size that is found in most insects, so we must look for a satisfactory explanation of the presumed advantage of producing variable-sized adults, particularly since extremes in adult size could easily present problems in mating. Aside from the physical handicaps associated with the copulation of different-sized mantispid individuals, small representatives of either sex might risk becoming a meal instead of a mate when approaching a larger individual. Thus, in consideration of this problem alone, it would seem that selection should favor size equality of the sexes, as it seems to in most insects, so that adults of smaller, but of nearly uniform, size might be expected from egg sacs regardless of the size of these sacs.

Every female must divide the energy available for reproduction between that utilized in positioning eggs in the environment and the energy actually contained in the eggs themselves. Time may well be the limiting factor for a female insect that spends a great deal of energy in locating sites, such as host animals or plants, for oviposition. For such a female there is no advantage in being able to produce more eggs than she has time to lay. But there should be an advantage in increased egg-laying capacity for a female that invests primarily in the eggs themselves.

Although we are ignorant of exactly where female *M. uhleri* lay their eggs, it seems likely that they expend little energy in positioning eggs. This conclusion is derived from the very large size of the egg clutches of *M. uhleri*, which indicates a low probability of larval success in locating spider eggs and that the larvae are pretty much on their own in locating food. Thus, the number of offspring a female can produce is not likely to be limited by time, but rather by the number of eggs she can physically synthesize, which is probably correlated with her size. Large size, whenever this is allowed by a surfeit of available spider eggs, should be valuable in species such as *M. uhleri* where the low probability of larval survival must be compensated for by the production of large numbers of eggs.

Of course, such arguments do not explain the advantage of large size to the male. This feature in the male may simply be carried along by selection in the female. Or perhaps, faced with a population of variable-sized females, large size in the male is valuable because an increase in size decreases the likelihood of being eaten by a larger female during courtship. It would seem advantageous for a male to mate with as large a female as possible, since this would increase his reproductive potential, and this may be more easily accomplished if the male is as large as his larval food supply permits.

Possible male pheromone

Eltringham (1932) postulated the existence of a male phero-mone as the product of an eversible glandular sac between the fourth and fifth abdominal segments of *M. styriaca*. An analogous situation occurs in *M. uhleri* since, while there is no eversible sac, there are porous areas with associated glandular epithelia in the flexible cuticle between the third and fourth, fourth and fifth, and fifth and sixth abdominal tergites of the male (MacLeod, unpub-lished). These areas are likely sources for the sweetish odor which can be detected during courtship. The observed flicking of the male abdomen and extension of his abdominal segments may serve to broadcast this pheromone. What seems to be this same odor has been apparent to us when examining freshly thawed specimens that had been preserved by freezing. Closer examina-tion of these specimens under a dissecting microscope has revealed small liquid droplets exuding from these porous areas of the abdomen.

3. Larval Strategies for Locating Spider-Egg Prey

The means by which a searching first instar mantispid larva locates a spider egg has remained an intriguing question for more than a century after Brauer (1869) demonstrated that larvae of *M. styriaca* can burrow directly through the wall of a spider egg sac. The reports of Hungerford (1939), Kaston (1938, 1940), Viets (1941) and Lucchese (1955, 1956) suggest that the larvae of their respective species board spiders; McKeown and Mincham's (1948) study, like Brauer's, confirms the findings of direct penetration of egg sacs by mantispid larvae.

The seeming contradictions in these observations are magnified by the implication that each investigator is presenting a fragment of evidence of some general tactic common to all larval mantispids. We contend that they can better be identified as two rather distinct strategies: the direct penetration of spider egg sacs that have been spun and left in the environment, and the mantispids' boarding of spiders prior to egg sac production and entrance into the sac at the time of spinning. We shall also argue that these two maneuvers may not necessarily be mutually exclusive in a particular species of mantispid, different mantispid species utilizing either tactic, or both, to varying extents.

Previous studies have been hampered by reliance on random observation rather than upon data derived in significant quantities from controlled situations. Negative findings are particularly difficult to evaluate, since, for example, if a larva of a particular species fails to penetrate an egg sac it is possible that the type and condition of the sac or its presentation may account for the failure of the larve to penetrate.

We have circumvented this problem and others through a series of experimental studies in which such variables as larval age and

25

species of spider producing the egg sac or presented for boarding were controlled. Further, we have employed as the experimental subjects larvae of *Mantispa viridis* and *M. uhleri*. Because of their rather different behaviors, detailed below, each species acts as a partial control for the results obtained from the other.

Our investigations of these strategies are dealt with in Experiments 2, 3, and 4. In Experiment 2, larvae of both species were exposed simultaneously to the same spider to see how their behavior differed toward it. In Experiment 3, the behavior of the two species toward egg sacs was investigated in a situation in which the surface area to be searched was an experimental variable. The activities within the egg sac and the developmental rates of the larvae of *M. uhleri* and *M. viridis* were compared in Experiment 4, and the important differences found between these are discussed relative to the distinctions in their methods of finding spider eggs.

Our results suggest that *M. viridis* is an obligate egg sac penetrator only, while *M. uhleri* is capable of this same penetrating behavior as well as the tactic of boarding spiders prior to egg production, entering the sacs subsequently as these are spun. These two strategies can be related to the findings of previous workers on other mantispid species.

METHODS AND RESULTS

First instar larvae of both *M. uhleri* and *M. viridis* were obtained from adults reared in the laboratory using the techniques already described (pp. 6-11). The culture of *M. viridis* was descended from females collected at Little Grand Canyon, Jackson Co., Illinois. Under low magnification, larvae of the two mantispid species were easily distinguished by the pattern of abdominal pigmentation, which in *M. uhleri* is composed of a series of transverse bands and in *M. viridis* consists of a bilateral pair of dorsal longitudinal stripes extending the length of the abdomen. All experiments were conducted with larvae that had hatched from the egg on the day on which the experiment was begun.

Experiment 2: Reaction of *M. viridis* and *M. uhleri* larvae to spiders

Spiders of several species (Table 6) representing five different families were collected. With the exception of the specimen of *Admestina tibialis* (C. L. Koch), collected at Carbondale, Illinois, all were taken in the area of Urbana, Illinois. Each spider was placed in a separate screw-top jar measuring 8.9 cm high and 7.2 cm in diameter with an internal surface of 233 cm². Holes had been drilled in the Bakelite jar tops for ventilation, and a circular piece of filter paper placed between the top and lip to prevent the escape of larvae. At the beginning of each exposure, 20 first instar larvae of *Mantispa viridis* and 20 of *M. uhleri* were released on the filter paper-lined jar top with a spider situated in the bottom of the jar. The jars were then left undisturbed for 24 hours, after

Table 6. Boarding frequency of *Mantispa uhleri* and *Mantispa viridis* on various spider species (Exper. 2)

Spider species	Number and sex used	Larvae boarding spider	
		M. uhleri	*M. viridis*
Dysderidae			
Ariadna bicolor	5 ♀	26/100[a]	1/100
Agelenidae			
Agelenopsis sp.	1 immature	5/20	0/20
Lycosidae			
Schizocosa sp.	3 immature	17/60	0/60
Pardosa milvina	2 ♂	6/40	0/40
Thomisidae			
Philodromus vulgaris	3 subadult ♂	40/60	0/60
Salticidae			
Phidippus audax	3 subadult ♀	40/60	0/60
Metacyrba undata	3 (♀, ♂, immature)	20/60	0/60
Admestina tibialis	1 ♀	1/20	0/20
Total larvae on spiders/total larvae presented to spiders		155/420	1/420

[a] 26 larvae boarded out of a total of 100 larvae presented to the spiders.

which the spiders were quickly placed in a Syracuse dish containing 70% ethyl alcohol. Spiders were then scored as to the number of larvae found either still attached or in the alcohol.

Of the 420 larvae of each species exposed to 21 spiders, 155 *M. uhleri* and 1 *M. viridis* were found associated with a spider (Table 6). Most of the *M. uhleri* were still on the spider, with only a few loose in the alcohol of the dish. These larvae most frequently took up a position on the pedicel of the spider, although larvae were also found between the spinnerets and around the bases of the legs. The single larvae of *M. viridis* was found loose in the alcohol of the dish.

Experiment 3: Behavior of *M. viridis* and *M. uhleri* larvae toward spider egg sacs

The egg sacs used were those of an agelenid, *Agelenopsis* sp., collected during the winter months in and around Urbana, Illinois, from beneath the loose bark of Osage orange trees. For laboratory rearings, we have collected overwintering *Agelenopsis* egg sacs from these same sites for the past five years and have never found any naturally occurring mantispid larvae within them; it is thus virtually certain that the larvae found within them at the conclusion of our investigations were our experimental larvae. The egg sacs are lenticular in shape and usually constructed flat against the inner surface of the bark and covered with a cone of loose silk and debris. Sacs were carefully separated from this cone before use in the experiment.

Two sets of three screw-top jars of increasing total surface were used: 135 cm^2 (6.4 cm high, 5.9 cm diam); 233 cm^2 (jar used in the preceding experiment); and 573 cm^2 (17.1 cm high, 9.5 cm diam). One *Agelenopsis* sac was placed in each of the six jars. At the beginning of each replication five newly hatched larvae of *Mantispa viridis* were placed in each of the three jars of one set, on the filter paper-lined Bakelite tops that sealed the jar; five *M. uhleri* larvae were placed in each of the three jars of the other set. Jars were left undisturbed for 24 hours after which they were opened and the egg sacs scored for larvae in and on the sac. Ten replications were run. In each case, all larvae were accounted for to ensure that none had escaped from the jar. The independence of

Table 7. *Mantispa uhleri* and *Mantispa viridis* penetration of *Agelenopsis* sp. egg sacs in controlled surface areas (Exper. 3)

Surface area of chambers	Number of *M. uhleri* larvae (N=50)		Number of *M. viridis* larvae (N=50)	
	Associated with sac	Not associated with sac	Associated with sac	Not associated with sac
135 cm²	19[a,c]	31	46[b,c]	4
233 cm²	9[a,d]	41	38[b,d]	12
573 cm²	4[a,e]	46	44[b,e]	6

[a] Frequencies of larvae associated with sac and not associated with sac differ significantly among three chambers ($X^2 = 13.90$, $P < 0.05$, df = 2).
[b] Frequencies of larvae associated with sac and not associated with sac not significantly different among three chambers ($X^2 = 5.54$, $0.1 > P > 0.05$, df = 2).
[c] Frequencies of larvae associated with sac and not associated with sac differ significantly between species in small chamber (X^2 adj = 29.71, $P < 0.005$, df = 1).
[d] Frequencies of larvae associated with sac and not associated with sac differ significantly between species in medium chamber (X^2 adj = 31.47, $P < 0.005$, df = 1).
[e] Frequencies of larvae associated with sac and not associated with sac differ significantly between species in large chamber (X^2 adj = 60.94, $P < 0.005$, df = 1).

the effect of surface area of the jar to be searched and mantispid species on the location of larvae was tested using the chi-square distribution.

Significantly more larvae of *M. viridis* than of *M. uhleri* were associated with the egg sac for all jar sizes (Table 7). The frequency of *M. uhleri* larval association decreased with increasing jar size with the difference among the three sizes being significant. The corresponding frequencies for *M. viridis* were not significantly different.

Experiment 4: Developmental Rates of *M. viridis* and *M. uhleri*

Larvae of both species were reared with use of *Achaearanea tepidariorum* eggs in pseudosacs covered by glass slides. Larvae were left undisturbed except for visual observations through the glass until the third instar, at which time enough spider eggs were added so that an excess remained at the time of cocoon spinning, ensuring that the duration of this stage reflected the larva's ontogenetic program rather than the amount of food available. Differences in the duration of corresponding stages in the two

Table 8. Duration of developmental stages of *Mantispa uhleri* and *Mantispa viridis*[†] (Exper. 4)

Developmental Stage	Mean days ± SE		t-test (two-tailed) comparison between species
	M.uhleri (N = 11)	*M. viridis* (N = 14)	
First instar	7.4 ± 0.3	4.4 ± 0.1	t = 10.10***
Second instar	2.3 ± 0.1	1.9 ± 0.1	t = 2.32*
Third instar	2.5 ± 0.2	2.0 ± 0.0	t = 3.28**
Prepupa (cocoon spinning to pupation)	5.7 ± 0.1	5.1 ± 0.1	t = 3.45**
Pupa	9.9 ± 0.2	9.0 ± 0.2	t = 3.63**
Total	27.7 ± 0.4	22.4 ± 0.2	t = 12.00***

[†] Fed eggs of *Achaearanea tepidariorum* at 25° C, L:D = 16:8, relative humidity = 80%.
* = 0.05 > P > 0.01.
** = 0.01 > P > 0.001.
*** = P < 0.001.

species were compared using a t-test. The durations of the developmental stages for *M. viridis* were significantly shorter than those for *M. uhleri* (Table 8).

DISCUSSION

Our first insight into the possibility that different species of mantispids might have basically different ways of reaching their spider-egg prey came while rearing mantispids in the laboratory and after placing a hatching egg clutch of *M. uhleri* and one of *M. viridis* in a large jar with a female jumping spider, *Phidippus audax*. After 24 hours the spider had spun a flimsy retreat around herself, and, with the naked eye, we could see hundreds of larvae milling around and over her body. The larvae of these species are easily distinguished under the magnification of a dissecting microscope and an examination revealed that the only larvae that had actually boarded the spider were *Mantispa uhleri*, while the larvae milling around her and oriented in close proximity to the silken retreat were all *M. viridis*.

It is likely that *M. viridis* larvae will only seldom be found on spiders (Table 6). Because each of the spiders used in this experiment represented a portion of the surface area available to mantispid larvae of both species, one might expect to count some larvae as having boarded that were simply walking across the surface of the spider at the instant that it was removed from the jar, and this is probably the explanation for the one *M. viridis* associated with *Ariadna bicolor* (Hentz). Although it is possible that *Mantispa viridis* is a spider boarder specific for families or species we did not test, this seems unlikely, in view of the apparently nonspecific nature of its egg sac-penetrating behavior. *Achaearanea* and *Agelenopsis* sacs were the first presented to *Mantispa viridis* larvae, and both were readily entered. Later, egg sacs from the jumping spider, *Metacyrba undata* (De Geer), were made available to searching *Mantispa viridis* larvae, and these, likewise, were penetrated. Such impartial penetrating and feeding behavior is not consistent with a narrow boarding range.

Larvae of *M. uhleri* have boarded every spider presented to them in the laboratory, including hunting spiders of the genera *Herpyllus, Lycosa, Phidippus,* and *Salticus,* as well as the web spinners *Achaearanea, Araneus,* and *Argiope,* provided that these latter were prevented from suspending themselves in a web. However, the natural prey range is probably somewhat narrower. The rather high incidence of spider boarding indicated by these data suggests that the location of spiders by *Mantispa uhleri* may not be random.

Although both species exhibit sac-locating and sac-penetrating behavior, *M. viridis* is more successful at this than *M. uhleri* (Table 7). An increase in the surface area on which the sac was located did not seem to affect *M. viridis,* since there were no significant differences in the numbers of larvae penetrating sacs in the three different-sized jars. However, the differences were significant in *M. uhleri.* The decreased efficiency in locating egg sacs in successively larger jars suggests that the location of egg sacs by this species is more nearly random.

These data are consistent with the hypothesis that *M. uhleri* encounters sacs through random searching, while larvae of *M. viridis* are able to orient to egg sacs from a distance. Furthermore, the data are also consistent with our added suggestion that some element of *M. uhleri*'s search for spiders may not be random, which leads to the prediction that *M. uhleri* differs importantly from *M. viridis* in actively orienting its search behavior toward spiders and directly penetrating sacs only if these are accidentally encountered along the way. Before this idea can be accepted, however, certain complicating factors must be eliminated. For example, larvae of *M. uhleri* may show strong orientation toward egg sacs and penetrating behavior after some specific amount of time has passed without encountering a spider. The area searched by the larvae of the two species in a given amount of time might also have its own significance. Quite obviously, a closer examination of such possibilities is needed.

The rates of development of these two species are very likely related to the differences in the routes by which these mantispids reach spider eggs. Thus, the more rapid development of *M. viridis*, especially in the first instar, should be adaptive for an egg sac penetrator that will probably encounter eggs in varying stages of development and occasionally have to contend with hatching spiderlings. But larvae of *M. uhleri*, entering the sac during its construction, from a position on the spider, could afford to feed in a comparatively leisurely fashion. Other behavioral aspects parallel the developmental rate. Third instar *M. uhleri* are extremely sluggish and can be damaged and killed by hatching spiderlings. Disturbing these larvae by touching them produces, at most, slight and extremely slow reactive movements. In contrast, third instar larvae of *M. viridis* are quite active and whip violently away when touched, a possible protective mechanism against hatching spiderlings.

On the basis of the preceding considerations we should like to suggest that, despite the fragmentary nature of some of the observations, all of the other mantispid species studied by previous workers also show one of the two basic strategies demonstrated in these experiments (Table 9). Although Brauer (1869) did not expose larvae of *M. styriaca* to spiders, and McKeown and Mincham's (1948) observations of *M. vittata* and spiders were not extensive, we would not expect to see spider-boarding activity in either of these species because of their documented overwintering as unfed first instar larvae. Spiders present for boarding in the spring would also have been available the preceding fall and would seem to offer a more protected overwintering site than the one actually used, which in *M. vittata* is an aggregation of larvae in a mass beneath loose bark. Larvae of

Table 9. Method of egg sac entry by various mantispids, as extrapolated from the literature

Mantispid species	Reference	Spider boarding	Egg sac penetration
Mantispa interrupta Say	Viets, 1940	+	?
Mantispa styriaca Poda	Brauer, 1869	–	+
Mantispa uhleri Banks	Present report	+	+
Mantispa viridis Walker	Present report	–	+
Mantispa vittata Guer.	McKeown and Mincham, 1948	–	+
Perlamantispa perla (Pallas)	Lucchese, 1956	+	?

Perlamantispa perla (Lucchese, 1956) and *Mantispa uhleri* do, in fact, overwinter on spiders. In the case of *M. uhleri* this overwintering behavior provides not only protection and insulation, but food as well, since the larvae actually feed on spider blood during this time (Redborg and MacLeod, 1983b). Thus, for a spider-boarding mantispid, there would be little advantage in waiting until spring to search for spiders.

Viets's observations are not adequate to rule out egg sac penetration for *M. interrupta*. Only one example of negative evidence is given, and the conditions under which the sacs were presented are not described in detail. Lucchese's findings that egg sac penetration by *Perlamantispa perla* does not occur are more convincing. Still, he is not specific about controls for his observations. It is not known, for example, whether those larvae that ignored egg sacs were simultaneously exposed to spiders of the same species that they did, in fact, board. In addition, Lucchese used large numbers of larvae in his investigations, a situation that might reduce the likelihood of sac penetration in favor of spider boarding in a species that does both. It does seem likely to us, however, that mantispids will be found which specialize in spider-boarding behavior to the exclusion of any direct penetration of egg sacs. *Perlamantispa perla* may well be one of these.

The clarification of these two rather different methods of locating spider eggs should help to illuminate such other aspects of mantispid biology as the possible choice of oviposition sites by females and the sensory cues used by searching first instar larvae.

4. Egg Sac Penetration

In our experiments the mantispid larvae were confined for long periods (ca. 24 hrs) with spiders or egg sacs, and were not allowed to leave the area; this of course is contrary to natural conditions. It could thus be argued that either egg sac penetration or spider boarding is a laboratory artifact brought about by the repeated, prolonged exposure of larvae to egg sacs or spiders.

The boarding of spiders can be confirmed in nature by examining wild-caught spiders for the presence of mantispid larvae. It is difficult, however, to secure data showing that in nature larvae of *M. uhleri* directly penetrate egg sacs. Collected in the field, an egg sac containing a larva tells us nothing of the larva's method of entrance. Direct penetration under laboratory conditions might be only a small facultative component of a much more elaborate sequence of behaviors following the obligate boarding of a female spider. A larva may ideally reach the eggs as they are being deposited. Direct egg sac penetration might be employed if and only if the larva reaches the vicinity of the eggs after the sac has been partially constructed.

In a series of six experiments we have examined some of the details of egg sac-penetrating behavior. The problem of whether direct egg sac penetration is normal larval activity or an artifact was addressed directly in Experiments 5, 6, and 7. Experiment 8 is concerned with some of the factors that affect egg sac penetration. Finally, Experiments 9 and 10 focus on problems encountered by larvae after egg sac penetration.

METHODS AND RESULTS

Egg sacs used in the following experiments were either those of *Achaearanea tepidariorum*, obtained as soon as spun from laboratory maintained, field-collected adults, or those of *Agele-*

nopsis sp. collected from beneath the loose bark of Osage orange trees and prepared as detailed for Experiment 3 (p. 28). We used adult crab spiders, *Philodromus vulgaris* Hentz, taken from the same habitats as the *Agelenopsis* sacs. All three spider species were collected in the vicinity of Urbana, Illinois. All experiments were carried out at a temperature of 25°C with 80% relative humidity and a photoperiod of L:D = 16:8. Frequencies were analyzed using the chi-square (X^2) distribution or the G statistic. Means were compared by a t-test.

Experiment 5: Discovery and penetration of egg sacs by unrestrained larvae

An experimental arena was constructed with an inverted 11.5-cm diameter Petri dish top, as illustrated in Figure 8. *Philodromus* spiders or *Agelenopsis* egg sacs, or both, were placed at positions P1 through P6 of Figure 8. Spiders were restrained by folding a small piece of cheesecloth in half, placing the spider between the two pieces thus formed, and closing the three open

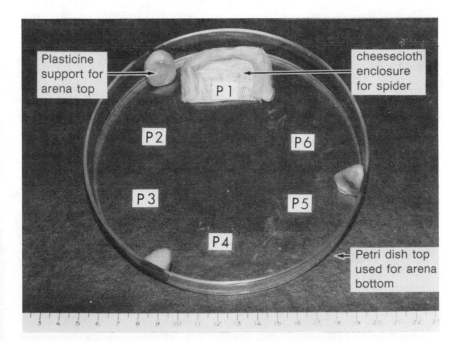

Fig. 8. Arena used in Experiment 5.

sides with Plasticine. Crab spiders were specifically chosen because of their inability to escape from such a restraint. Eggs or larvae of *Mantispa uhleri* were positioned in the center of the arena at the beginning of each experimental run. The arena was covered by the inverted bottom of the Petri dish, supported by three pieces of Plasticine; the wide gap thus created between the arena floor (Petri dish top) and arena covering (Petri dish bottom) allowed larvae to leave the arena freely around its circumference. Since the underside of the arena bottom was not completely flat, it was placed on a large sheet of cotton wool to ensure even contact between the bottom and the surface on which the arena was placed so that larvae would not become trapped on the surface of the arena. The experiment was run four times, with different combinations of spiders and egg sacs as follows:

Run 1. Spiders were placed at positions P1, P3, and P5 and egg sacs were placed at positions P2, P4, and P6. Egg sacs were placed in cheesecloth enclosures identical to those used for the spiders. Just prior to hatching, one entire *M. uhleri* egg clutch on filter paper was placed in the center of the arena.

Run 2. Egg sacs were placed at all six positions with only those at P1, P3, and P5 being placed in cheesecloth enclosures. Just prior to hatching, an entire mantispid egg clutch on filter paper was again placed in the center of the arena.

Run 3. Spider egg sacs were placed at all six positions. No cheesecloth enclosures were used. A group of 3-day-old mantispid larvae hatched from a single egg clutch was allowed to disperse from the center of the arena. These larvae had been stored in a cotton-stoppered 2-dram shell vial since hatching. The vial was opened and placed vertically in the center of the arena at the beginning of the run.

Run 4. Egg sacs were placed at all six positions without cheesecloth enclosures. One hundred newly hatched first instar larvae from the same clutch were released into the arena, 20 at a time, over a 3-day period. The clutch was kept in a stoppered shell vial and larvae were individually transferred with a small camel's hair brush to a small piece of filter paper that was placed in the center of the arena. Each increment of 20 larvae was released only when all larvae from the previous release had dispersed and were no longer visible.

All runs were scored for the number of larvae found on spiders or in egg sacs. Under CO_2 anesthesia, spiders and egg sacs were examined through a dissecting microscope. Empty egg shells from the clutches used in the first three runs were counted to determine the number of larvae available in the arena. Results are summarized in Table 10.

Table 10. Pentration of *Agelenopsis* sp. egg sacs by unrestrained first instar *Mantispa uhleri* (Exper. 5)

Run	Available larvae	Larvae on spiders	Larvae in egg sacs
1	1,398	72[a]	0[a]
2	1,087	not used	3[b, d]
3	842	not used	8[b, c]
4	100	not used	4[c, d]

[a] Frequencies of spider boarding versus sac penetration significantly different (χ^2_{adj} = 71.864, P < 0.001, df = 1).
[b] Frequencies of sac penetration between Runs 2 and 3 not significantly different (G_{adj} = 2.722, 0.1 > P > 0.05, df = 1).
[c] Frequencies of sac penetration between Runs 3 and 4 not significantly different (G_{adj} = 3.195, 0.1 > P > 0.05, df = 1).
[d] Frequencies of sac penetration between Runs 2 and 4 significantly different (G_{adj} = 8.330, 0.005 > P > 0.001, df = 1).

Experiment 6: Direct observation of egg sac penetration by larvae in a confined area

A 27-cm^2 arena was constructed by filling the inside of a 5.9-cm diameter Bakelite jar lid with Plasticine, to weight it down, and placing it right side up into a Petri dish of larger diameter (Fig. 9). Water was poured into the Petri dish until it just reached the top of the Bakelite lid, forming a circular surface surrounded by water that effectively isolated a first instar larva on an island. *Achaearanea* egg sacs were utilized in this experiment, one sac being placed in the center of the arena at the beginning of each run. One first instar *Mantispa uhleri* was placed at the circumference of the arena with a fine camel's hair brush. The larvae used in this experiment varied in age from 1 to 4 days. The different lots were not siblings. Movements of the larvae were observed and times were recorded for the following events to occur: (1) egg sac hit (a hit was recorded when a larva disappeared from sight beneath the sac viewed dorsally); (2) egg sac mounting; (3) initiation of egg sac penetration (initiation was recorded when walking ceased and

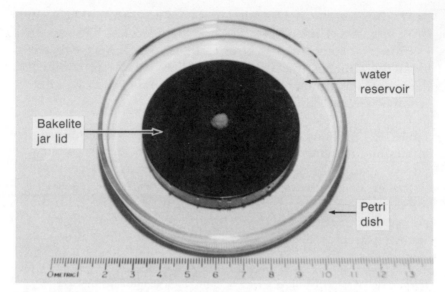

Fig. 9. Arena used for observing larvae in Experiment 6.

side-to-side movements of the head capsule against the silk were observed); (4) time interval required to penetrate an egg sac and completely disappear from view. The behavior of 12 larvae was recorded (Table 11), their movements observed through a dissect-

Table 11. Behavior of twelve first instar larvae of *Mantispa uhleri* on egg sacs of *Achaearanea* (Exper. 6)

Larva	Number of encounters with sac	Mounting of sac	Minutes Time until mounting	Minutes Time from mounting to penetration
1	1	+	14	18
2	1	+	14	12
3	1	+	7	17
4	1	+	2	6
5	1	+	5	34
6	1	−	−	−
7	2	+	7	20
8	2	+	6	10
9	2	+	6	12
10	3	+	32	9
11	4	+	12	52
12	4	+	16	38

ing microscope at 10X magnification. Observations were terminated after a period of 45 minutes if the larva had not yet mounted the egg sac. Only 1 in 12 larvae failed to mount and subsequently penetrate the egg sac within the allotted period.

Experiment 7: Larval feeding responses to real sacs and pseudosacs

Pairs of *Achaearanea* egg sacs of the same age were collected and stored at 10°C until used. One sac of each pair was randomly selected, opened, and its contents placed in a standard pseudosac. The other sac of the pair was placed unopened in a 2-dram shell vial containing in the bottom ¼ inch of a hardened mixture of plaster of Paris and activated carbon, the same material used for the pseudosac. Two first instar larvae on their day of hatching from the same egg clutch were used in each run. With a fine camel's hair brush one larva was placed on the eggs in the pseudosac which was then closed by wetting the area around it with distilled water and gently pressing a small piece of standard 1-mm thick microscope slide over the opening until a seal was made. The second larva was placed on the *Achaearanea* sac in the vial and a piece of cotton wrapped in a Kimwipes was then pushed into the vial until it just touched the egg sac. The pairs were scored as to the final developmental stage reached by each larva, and, if an adult was produced, the length of time taken until eclosion was noted (Table 12). No significant differences were observed between the two groups.

All mortality beyond the first larval stadium occurred in egg

Table 12. Behavior of first instar *Mantispa uhleri* larvae toward *Achaearanea* spider eggs in real egg sacs and pseudosacs (Exper. 7)

Eggs in	Total larvae exposed to eggs	Number not feeding	Number dying during development	Number of adults produced	Days to adult emergence (Mean ± SE)
Pseudosacs	44	2[a]	11[a]	31[a]	24.2 ± 0.19[b]
Real sacs	44	6[a]	9[a]	29[a]	24.4 ± 0.20[b]

[a] Frequency distributions between pseudosacs and real sacs not significantly different ($\chi^2 = 2.267$, $0.75 > P > 0.50$, df $= 2$).
[b] Means not significantly different (t $= 0.723$, $0.5 > P > 0.2$, 2-tailed).

---

sacs from which spiderlings hatched before larval development was complete. In most instances black dots and blotches could be observed on the larval cuticle.

Experiment 8: Larval age and egg sac integrity in relation to sac-penetrating behavior

The same type of medium-sized jars described in Experiment 3 were used. Each jar contained one *Agelenopsis* egg sac, situated on the jar bottom. At the beginning of each replication, five first instar *Mantispa uhleri* larvae were released on the filter paper-lined top of each jar. After 24 hours each egg sac was scored for the number of larvae inside. The experiment was run three times, each with 10 replications. In Run 1, freshly hatched larvae and completely closed egg sacs were used. In Run 2, newly hatched larvae from the same clutch as Run 1 were used, but the *Agelenopsis* sacs were slightly torn so that an open path to the eggs existed. Run 3 was identical to Run 1, using 2-day-old larvae from the same clutch.

Results (Table 13) indicate that neither larval age nor egg sac integrity is a significant factor in the mantispid's egg sac-penetrating behavior.

Table 13. Effect of larval age and egg sac integrity on penetration frequency of *Agelenopsis* egg sacs by first instar *Mantispa uhleri* (Exper. 8)

Location of larvae	Number of larvae			
	0-1-day-old		2-3-day-old	
	Whole sac	Torn sac	Whole sac	Torn sac
In egg sac[a]	4	6	8	–
Not in egg sac	46	44	42	–

[a] Frequencies of larvae in egg sacs not significantly different among three groups ($\chi^2 = 1.515$, $0.5 > P > 0.1$, df = 2).

Experiment 9: Developmental inhibition related to number of larvae per sac

For each replication three *Achaearanea* egg sacs of the same age were opened and their contents emptied into each of three standard pseudosacs. Freshly hatched *Mantispa uhleri* larvae from the same clutch were placed in the three pseudosacs as follows: One larva in sac #1 (Group 1); two larvae in sac #2 (Group 2); and three larvae in sac #3 (Group 3). Larvae were inserted and the pseudosacs closed as described in Experiment 7. No eggs were added. Under our visual monitoring each pseudosac was left undisturbed until a larva reached the spinning stage or until hatching spiderlings were evident within. The time elapsed and the stage reached by each larva were recorded (Table 14). Fourteen replications were made.

The larvae dying during the first stadium in all three groups showed no signs of feeding. The larva dying during the second stadium in Group 2 and four of the dead third instar larvae and the dead pupa in Group 3 died in association with hatching spiderlings. These dead larvae also showed black discolorations on their cuticles. The dead second instar larva and the remaining two unsuccessful third instar larvae in Group 3 had been killed and eaten by another larva. The frequency of larval survival past the first stadium decreased through the three groups with the differences among them significant ($X^2 = 8.810$, $.025 > P > .01$, df = 2). The number of days to cocoon spinning increased significantly from Group 1 to Group 2 (t = 3.518, $0.001 > P > 0.0005$, 1-tailed) and from Group 2 to Group 3 (t = 2.005, $0.05 > P > 0.025$, 1-tailed). In no instance did more than one larva reach maturity in any given pseudosac.

Table 14. Survival of competing larvae in an egg sac (Exper. 9)

Larvae per pseudosac	Number of pseudosacs	Number of mantispids					Number of larvae surviving past first instar/total	Days to cocoon spinning (Mean ± SE)
		Final stage reached by larva			Pupa	Adult		
		Larval instar						
		1	2	3				
1	14	2	0	0	0	12	12/14	10.0 ± 0.21 (N = 12)
2	14	15	1	0	0	12	13/28	11.5 ± 0.38 (N = 12)
3	14	25	1	6	1	9	17/42	12.7 ± 0.49 (N = 11)

Experiment 10: Egg sac age and larval development

Ten *Achaearanea* egg sacs from laboratory-maintained females were collected on each of four consecutive days and stored at 25°C. The experiment was begun the last day of collection, with the 40 egg sacs in four age groups as follows: Group 1 (0-24 hours old); Group 2 (24-48 hours old); Group 3 (48-72 hours old); and, Group 4 (72-96 hours old). Each sac was placed in a 2-dram shell vial at the bottom of which was ¼ inch of a hardened mixture of plaster of Paris and activated carbon moistened to raise the humidity. All larvae used were siblings from a single egg clutch. One freshly hatched *Mantispa uhleri* larva was placed on each sac, and a Kimwipes-wrapped cotton plug was inserted into the end of each vial until it just touched the egg sac. The sacs were left undisturbed until the emergence of an adult mantispid or spiderlings. If spiderlings emerged, the sac was examined to determine the stage reached by the larva.

The number of larvae successfully reaching the adult stage decreased as the age of the egg sac increased. Larvae that did not develop beyond the first stadium failed either to penetrate their sacs or to begin feeding. Larvae dying in the second or third stadia had entered sacs that produced hatching spiderlings before the larvae had reached the cocoon-spinning stage. As noted for Experiments 7 and 9, the cuticles of most of these dead second and third instars exhibited black marks. Results are summarized in Table 15.

Table 15. Effect of age of *Achaearanea* egg sacs on the development of *Mantispa uhleri* larvae[a] (Exper. 10)

Group/Age of eggs (days)	Number of sacs	Number of mantispids					
			Stage reached by larvae after penetration				
		Failing to penetrate	Larval instar			Pupa	Adult
			1	2	3		
1/0-1	10	0	1	0	2	0	7
2/1-2	10	2	0	0	3	0	5
3/2-3	10	0	0	3	6	0	1
4/3-4	10	1	2	6	1	0	0

[a] Frequencies of larvae reaching the adult stage for Groups 1 and 2 combined vs. Groups 3 and 4 combined were significantly different ($\chi^2_{adj} = 11.396$, $P < 0.005$, df = 1). Groups were combined to avoid cells with expected frequencies less than 5.

DISCUSSION

Results of several experiments are consistent with the conclusion that direct egg sac penetration is a naturally occurring phenomenon. In Runs 2-4 of Experiment 5 a total of 15 larvae penetrated the *Agelenopsis* sacs. These larvae were not confined and had the opportunity of leaving the arena as an alternative to mounting and penetrating an egg sac. Where larvae were confined (Experiment 6), 11 of 12 penetrated an egg sac within 64 minutes, and 5 of these larvae entered immediately after their first encounter with the sac. These encounters were thus neither prolonged or repeated and gave no indication of being "artificial." If direct penetration is not a part of *Mantispa uhleri's* natural behavior, then the barrier presented by intact egg sacs (Experiment 7) might be expected to produce a decrease in survival or an increase in development time, compared to larvae placed directly with spider eggs. Such differences could possibly be explained in terms of the time wasted in a prolonged search for spiders and the final, inefficient penetration of the sac. The fact that no significant differences were found suggests that larvae penetrated quickly and efficiently. Our results do not bear out the notion that egg sacs are directly entered only after repeated encounters.

Like Pandora's Box, however, Experiment 5 suggests much more than was first anticipated. The information obtained in Run 1, in which both spiders and egg sacs were available and no larvae penetrated or were even found close to the egg sacs, was totally unexpected. In addition to the 72 larvae found on the spiders, a total of 154 dead larvae were found directly beneath the spiders. Presumably the spiders could accommodate only so many larvae, and since larvae feed on spider blood after boarding (Redborg and MacLeod, 1983b), these dead individuals may have starved while attempting to board or been killed by their successful competitors.

As indicated, this apparent orientation to the spiders was highly significant. This result might be accounted for either by factors increasing the likelihood of spider boarding or by factors decreasing the probability of egg sac penetration.

Factors increasing spider boarding might involve the ability of larvae to locate spiders from a distance, whereas egg sacs are encountered only by random search, or—if both spiders and sacs are located at random—a greater frequency of abandonment and continued searching after encountering an egg sac than after locating a spider. But the number of larvae penetrating in Runs 2, 3, and 4 where spiders were not present was surprisingly low, since direct observation (Table 11) indicates that any larva encountering an egg sac has a high probability of entering. This result suggests the existence of a factor that decreases the probability of egg sac penetration under the conditions of Experiment 5. A hypothesis consistent with all experiments is the existence of an inhibitory mechanism that prevents larvae from penetrating egg sacs located close to their hatching egg clutch, but which does not affect spider boarding. Such a mechanism could be adaptive since an egg sac located near hatching eggs of *M. uhleri* is likely to be encountered by many larvae. As detailed below, there is good evidence that only one larva will reach maturity in any egg sac. Thus, if two larvae penetrate an egg sac simultaneously, only one larva is likely to survive to cocoon spinning. Behavior discouraging a larva from penetrating a sac under such conditions should be favored by selection. This would not seem to be the case for the boarding of spiders which, unlike stationary egg sacs (excepting those of Lycosidae and Pisauridae which are carried by the egg-laying female), may move briefly through the vicinity of hatching larvae; a mobile spider temporarily located near a hatching mantispid egg clutch would not be as likely to pick up many larvae, compared to an egg sac. That an egg sac will be encountered by many larvae is assured. Thus, an appropriate reaction for a larva might be to ignore, or at least be wary of, egg sacs near its hatching site, but to board a spider, in any event.

There would seem to be two possible ways in which an inhibitory mechanism acting to prevent the penetration of nearby egg sacs might work. First, larvae might be unresponsive to egg sacs for a given amount of time, allowing them sufficient opportunity to disperse some distance from the egg clutch. The time might be measured by some sort of energy clock related to distance traveled. If this were the case, larval age would be expected to have considerable effect on penetration activity. In Run 3 (Table 10) where larvae were 3 days old, the number penetrating was not significatly greater than in Run 2 where freshly hatched larvae were used. This lack of any correlation with larval age is also corroborated by the results of Experiment 8 in which there was no apparent difference in egg sac-penetration frequency between 0-1 and 2-3 day-old larvae.

A second possibility is the inhibition of sac-penetrating activity by the presence of other larvae. As larvae disperse, the number of larvae per unit area would decrease until at some point penetrating behavior is activated. This mechanism would have the advantage of being self-adjusting for egg clutches of different sizes that are likely to occur because of *M. uhleri's* extreme adult size variation. By chance, the number of larvae hatching from the egg clutches decreased in Runs 1-3 and, by design, Run 4 had no more than 20 larvae in the arena at any particular time. Thus, the larval density at any given time was lower in each succeeding run. The number of larvae penetrating sacs increased from Runs 1-4, and the differences in the frequency of penetration between Runs 2 and 4 were significant. This partial result is consistent with the hypothesis just outlined. Very likely, additional possibilities accounting for these results could be proposed, and the existence of some such mechanism is worth further investigation.

One factor that could affect egg sac penetration is the texture and composition of the egg sac. The difficulty experienced by a larva of *M. uhleri* in penetrating an egg sac may conceivably vary with the species of spider producing the sac. *Achaearanea* egg sacs are penetrated in less than one hour. In this case, the larva crawls over the sac until it finds a suitable point of entry. The head, appressed to the surface of the sac, is then moved from side to side, appearing to abrade the surface. The larva enlarges the opening thus made, moves into the sac, and disappears beneath the silk. Although microscopy does not reveal any obvious adaptations in *Mantispa uhleri*, it is possible that the mouthparts may either cut the silk or release a matrix-dissolving enzyme. Closer study of the exact mechanism of penetration is required. The low rate of penetration of *Agelenopsis* sacs (Experiment 5) cannot be explained solely by difficulty of entrance, since sacs torn open to facilitate entry (Experiment 8) had no influence on the number of larvae entering sacs.

Once a sac is entered, a larva may encounter several problems, a major one being penetration by additional larvae. Our findings (Table 14) strongly suggest that no more than one *Mantispa uhleri* will reach maturity in any egg sac. Interestingly, it also appears that only one begins development, the other unsuccessful larvae dying (or being killed) without appreciable feeding. Indeed, the lack of feeding by the unsuccessful larvae suggests that the larvae actively search out one another and interact until just one remains alive. Only then does feeding begin. Of course, the possibility exists that the larvae simply refrain from feeding until starvation eliminates all but one, but it seems unlikely that a larva would passively starve with available food present. Although it might seem strange that larvae do not begin feeding immediately in an attempt to "outeat" competitors, there may be forces operating against this. Noteworthy is the fact that the larva that begins to feed at once will also be the first to reach the more vulnerable, quiescent intermolt period, during which assassination by a smaller, more agile larva might be comparatively easy. In all three instances in which two larvae began development, one was eventually killed by the other and its contents ingested. Further data derived from Experiment 9 (Table 14) support the hypothesis that larvae compete with each other in the egg sac and mutually inhibit development. Here the number of larvae beginning development decreased as the number of larvae per sac increased. The significant increase in time to cocoon spinning among the three groups is also consistent with the conclusion that as the number of larvae per sac increases, so does the time until elimination of all but one larva.

A second factor at work is the age of the egg sac. If the sac is too old, a larva may not have enough time to complete feeding and spin a cocoon before the spider eggs hatch (this would not happen to larvae entering freshly spun sacs after having first boarded the spider). This problem is documented in Table 15. Development in *Achaearanea* is quite rapid, with spiderlings hatching from the egg sac 11-12 days after construction at 25°C in the laboratory. As the age of the egg sac prior to its entry by a larva increases, the number of adult mantispids emerging decreases and the average final stage reached by unsuccessful larvae decreases. Thus, only a single adult was obtained from 20 egg sacs 2-4 days old, while 20 egg sacs 0-2 days old produced 12 adult mantispids. This difference was highly significant: seven of ten larvae developed to the third instar in Group 3, while only one of ten did so in Group 4.

These results indicate that unless a *Mantispa uhleri* larva locates an *Achaearanea* egg sac within 48 hours of its construction, the probability of the larva's developing to the adult is extremely low. Depending on the species of spider, the larval maneuver of direct egg sac penetration may be a rather risky affair. *Achaearanea*, for example, has such a short developmental period that the 11-12 days needed for the hatching of the spiderlings is only slightly longer than the time needed for development to cocoon spinning by larvae of *Mantispa uhleri*. But such rapid spider egg development may be the exception, and species such as *Lycosa rabida* Walckenaer and *Phidippus audax*, which need approximately 30 days in the laboratory to produce emerging spiderlings, would allow *Mantispa uhleri* a longer time for successful penetration.

It is interesting that larvae in these experiments penetrated and began feeding in sacs in which they were destined not to survive. It would be adaptive for mantispid larvae to be able to test the developmental state of eggs within an egg sac and to abandon sacs that would provide insufficient time for successful cocoon spinning. *Achaearanea's* rapid egg development might interfere with such an ability and it would be worth investigating a possible existence of such a mechanism in relation to spiders with a longer egg development time.

5. Boarding of Spiders

Our understanding of spider boarding has been limited to the supposition that it affords the boarding mantispid larva an opportunity to enter an egg sac while it is being produced. Larval behavior toward spiders prior to and after boarding has heretofore been a mystery.

In this chapter we examine factors affecting spider boarding in a series of three experiments in which first instar larvae of *M. uhleri* were exposed to spiders under various conditions and scored as to whether boarding occurred. The actual boarding process is described from observations of restrained spiders.

METHODS AND RESULTS

The following experiments were carried out at a temperature of 25°C and a photoperiod of L:D = 16:8. Frequencies of spider boarding in the various experiments were analyzed using the chi-square distribution.

Experiment 11: Spider sex and larval boarding

Mature males and females of the salticid *Metacyrba undata* were collected from overwintering retreats, usually beneath the loose bark of trees, in several Illinois localities. A series of three screw-top jars with increasing internal surface areas identical to those described in Experiment 3 were used. As in Experiment 3, small holes were drilled in the Bakelite jar tops for ventilation, and a piece of circular filter paper placed between the top and jar lip effectively confined larvae within the container. Five first instar larvae of *Mantispa uhleri* were placed on the filter paper

Table 16. Behavior of first instar *Mantispa uhleri* larvae boarding *Metacyrba undata* adults (Exper. 11)

Surface area of experimental chambers	Number of Larvae Boarding/ Available larvae		Combined data from ♀ and ♂ spiders
	♀ Spiders	♂ Spiders	
135 cm²	7/25	13/40	20/65 [b, d]
233 cm²	15/25	21/40	36/65 [b, c]
573 cm²	6/25	8/40	14/65 [c, d]
Totals	28/75 [a]	42/120 [a]	70/195

[a] Boarding frequencies (number of larvae boarding spiders versus number not boarding) of male and female spiders not significantly different (X^2_{adj} = 0.034, 0.9 > P > 0.5, df = 1). Data for three jar sizes were lumped for each sex. Homogeneity (X^2 = 0.109, df = 2), not significant, indicating that lumping was justified.
[b] Boarding frequencies significantly different (X^2_{adj} = 7.058, 0.01 > P > 0.005, df = 1).
[c] Boarding frequencies significantly different (X^2_{adj} = 14.332, P < 0.005, df = 1).
[d] Boarding frequencies not significantly different (X^2_{adj} = 0.996, 0.5 > P > 0.1, df = 1).

of the tops of each of the three jars at the beginning of each replication and one spider was placed in each jar. All spiders within a replication were the same sex. In eight replications male spiders were used and, in five, females were used. All fifteen larvae of any replication were siblings that had hatched the day of the experiment. After 24 hours, spiders were removed from the jars, dropped in a small dish of 70% ethyl alcohol, and scored for the number of larvae that had boarded.

A total of 42 out of 120 larvae boarded males of *Metacyrba undata*, while 28 of 75 larvae boarded females (Table 16). Larvae boarded either sex with equal frequency.

Experiment 12: Spider size and larval boarding

Forty-two immature salticids, *Phidippus audax*, were reared to the second (20), third (10), and fourth (12) stadia from eggs obtained from laboratory-maintained females. Spiders of each instar were placed individually in 2-dram vials with one first instar *Mantispa uhleri*. Two groups of second instar *Phidippus audax* were used: those that were newly emerged from the egg sac and active (10) and those that had fed for several days on *Drosophila melanogaster* Meigen and were partially engorged

(10). Vials were stoppered with cotton plugs wrapped with Kimwipes and placed in a humidifier at 80% relative humidity. They were examined daily until the mantispid boarded the spiderling, was eaten, or had died. Larvae eaten by spiderlings were easily identified by their shriveled and mutilated appearance. Larvae dying from other causes were dehydrated, but not mutilated.

Results are presented in Table 17. No larvae successfully boarded unfed second instar *Phidippus*. As spider size increased, so did the frequency of larvae successfully boarding.

Table 17. Immature *Phidippus audax* spiders boarded by first instar *Mantispa uhleri* larvae (Exper. 12)

Developmental stage of spider	Fate of mantispid larvae			
	Eaten	Boarding	Dead	Total
Second instar Unfed	10	0[a]	0	10
Second instar Engorged	5	5	0	10
Third instar Newly molted	5	5[a]	0	10
Fourth instar Newly molted	1	10[a]	1	12

[a]Boarding frequencies (number of mantispid larvae boarding spiders vs. number not boarding) for unfed second instar and newly molted third instar spiderlings combined (to avoid contingency table cells less than 5) and compared to boarding frequencies for newly molted fourth instars. Frequencies significantly different ($x^2_{adj} = 8.061$, $P < 0.005$, df = 1).

Experiment 13: Larval behavior toward previously boarded spiders

Naturally infested spiders of the species *Phidippus audax* and *Metacyrba undata*, each carrying one *Mantispa uhleri* first instar larva, were collected at the University of Illinois Dixon Springs Agricultural Center in southern Illinois. Each infested spider and an uninfested control spider of the same species, state of maturity, and sex (in the case of adults) were individually confined in 2-dram shell vials with one laboratory-hatched *M. uhleri* larva as described in Experiment 12. After 24 hours, spiders were anesthetized under CO_2 and scored as to whether boarding by the laboratory-reared larva had occurred. Since larvae on spiders slowly feed on spider blood (Redborg and MacLeod, 1983b), wild-caught larvae were easily distinguished from laboratory-reared larvae by their darkened midgut. Twelve pairs of spiders were tested during the period of several days necessary to collect them. Wild-caught larvae were either reared to the adult for identification or were identified by means of an unpublished key devised by the second author.

The frequencies of larvae boarding naturally infested spiders and uninfested controls were the same (Table 18). None of the wild-caught larvae moved from their original positions after the spiders were boarded by a laboratory-reared larva.

Table 18. *Mantispa uhleri* larvae boarding spiders already carrying a larva (Exper. 13)

	Number of Spiders	
Tested	Boarded	Not boarded
5 Infested immature male *Phidippus audax*	4	1
5 Uninfested controls	4	1
2 Infested immature female *Phidippus audax*	2	0
2 Uninfested controls	2	0
2 Infested immature *Metacyrba undata*	0	2
2 Unifested controls	0	2
3 Infested adult male *Metacyrba undata*	3	0
3 Uninfested controls	3	0
Total infested spiders	9	3
Total controls	9	3

Observations of spider boarding

A 27-cm^2 arena was constructed by filling the inside of a 5.9 cm diameter Bakelite jar lid with Plasticine to weight it down and placing it right side up in a Petri dish of larger diameter. Water could be poured into the Petri dish until it just reached the top of the Bakelite lid, forming a circular arena surrounded by water that could effectively confine a first instar larva. A

mature female of *Salticus scenicus* (Clerck) was restrained in the
center of the arena by applying a small drop of Elmer's Glue-all
over the tip of each leg while the spider was immobilized by CO_2
anesthesia. After the glue had dried the spider was allowed to
revive and the reservoir surrounding the arena was filled with
water. A single first instar larva *Mantispa uhleri* was placed at
the edge of the arena and its actions were noted through a
dissecting microscope. Four spiders were observed, each being
exposed, sequentially, to several different larvae. The movements
of a total of 20 larvae were observed.

During the initial stages of our observations on the boarding
behavior of larvae it proved impossible to restrain adult females
of such larger spider species as *Phidippus audax*, as the spiders
pulled free, and other means of restraint did not produce a
natural orientation of the spider. *Salticus scenicus* was then
chosen because of its small size. Although this species is not
known to be utilized by *Mantispa uhleri* in nature, we do not
believe that the observations arising from this experiment are
likely to be misleading since the range of spider species utilized
by *M. uhleri* is so broad. Adult females were used rather than
immature spiderlings or males, since, if larvae ever discriminate
between immature and adult, or between mature male and
female, we expect that it is the latter that should be preferred.

Direct observations of spider boarding revealed no discernible
pattern to the larva's movements before encountering the spider.
The larvae spent most of their time running about, circling the
water margin and periodically crossing the arena. After passing
beneath the spider's legs, however, each larva suddenly appeared
"alert." In a typical case the larva circled the spider, spending
most of its time in the area around the abdomen and near the
pedicel. Twitching of the spider's abdomen and flexion of its
legs were observed in response to larval contact. Boarding
occurred when the larva lifted its legs toward the spider's
abdomen, remaining attached to the substrate by its caudal
pygopod, and was whisked aboard the spider as the latter
brushed against the larva. In a series of short, jerky movements,
the larva made its way to the pedicel where it wrapped itself
around it like a belt. In all instances, the larva boarded within 30
minutes of its first encounter with the spider.

DISCUSSION

Mantispid larvae showed no clear-cut preference (Table 16) for male or female spiders in any of the three experimental jars. The larvae boarded both sexes, although, if given a choice, a preference might have been indicated. On the assumed equivalency of larval behavior toward male and female spiders, male and female data were combined and the numbers of larvae that boarded spiders in each of the three test jars were analyzed. A significantly greater number of larvae boarded spiders in the medium-sized jar. This unexpected finding suggests complexities in the larval searching-behavior that we do not yet understand. One possible explanation would involve the antagonistic effects of random larval searching (in which the number of boardings would decrease with increasing surface area) and larvae interfering with each other's boarding attempts (in which interference might decrease with increasing jar size, resulting in an increase in boarding).

First instar *Mantispa uhleri* are within the prey size range for second, third, and fourth instar *Phidippus audax* (Table 17). The number of larvae boarding spiders, increasing in correlation with spider instar and size, presumably reflects a decreasing desirability or visibility of the mantispid larva as a food item. We assume that there is a certain spider size above which first instar *Mantispa uhleri* are never consumed as prey. We observed that larvae are eaten by large spiders only after having been—by chance—picked off the spider's body by the grooming movements of the legs; these movements often were elicited in response to an apparent irritation caused by the larva. This seems to occur more commonly with long-legged spiders, such as lycosids, than with shorter-legged forms, such as salticids.

Our results also suggest the possibility that the nutritional state of the spider may affect the larva's success in boarding. Thus, while no unfed second instar *Phidippus audax* was boarded by a larva, half of the engorged second instar salticids were boarded.

On the basis of our previous evidence that only one larva will reach maturity in a given egg sac, it seemed conceivable that the presence of a resident *Mantispa uhleri* larva on a potential host spider might prevent a second larva from boarding, either by an inhibition of the boarding behavior of the second larva or by an aggressive action by the resident larva after boarding. Experimental results (Table 18) indicate that no such mechanisms are in operation. Consistent with this conclusion was the observation that no resident wild-caught larva changed its position after a subsequent boarding by a laboratory-reared larva. This would not have been expected if the first larva were aware of the second larva's boarding and had tried to prevent it. Although we know that a single larva on a spider will enter an egg sac at the time of oviposition, we have not seen reactions of multiple larvae on a spider at the time of egg production. Such multiple boardings do occur in nature. This phenomenon warrants further investigation.

6. Movements of First Instar Mantispid Larvae on Spiders

We now consider the behavior of a *Mantispa uhleri* larva after it has boarded a spider. Of particular interest are such questions as whether the larva can negotiate a spider molt; what is the preferred location on the spider's body adopted by a resident larva; whether, in fact, larvae on spiders are ultimately able to enter egg sacs from this position during oviposition by the spider; what happens to larvae that have boarded male spiders.

We have investigated these questions in a series of three laboratory experiments (14, 15, 16) with the salticid spider *Phidippus audax* and the lycosid *Lycosa rabida*, both of which are boarded by larvae of *Mantispa uhleri* in nature. In these experiments single first instar larvae were allowed to board immature spiders which were then reared—to the adult stage where possible.

METHODS AND RESULTS

All procedures were carried out at a photoperiod of L:D = 16:8 and a temperature of 25°C. The three experiments detailed below have certain features in common as they were carried out sequentially, with refinements added to Experiments 15 and 16 as we learned of the complexity of the phenomenon under study. The various activities in all experiments were analyzed in two-by-two contingency tables using the chi-square distribution or the Fisher Exact test.

Experiment 14: Larval behavior on nearly mature *Phidippus audax*

Large but immature *Phidippus audax* were collected in the vicinity of Urbana, Illinois, and at the University of Illinois Dixon Springs Agricultural Center (DSAC) in southern Illinois

55

during the late fall and winter. Each spider was placed in a 2-dram shell vial with one newly hatched first instar larva. Vials were stoppered with a cotton plug wrapped with Kimwipes and placed in an 80% relative humidity chamber. The spiders were examined at 24-hour intervals until larvae were no longer observed crawling in the vials, at which time each spider was examined under CO_2 anesthesia in order to note and record larval positions. Spiders were then returned to their respective shell vials and each was fed one house fly, *Musca domestica* Linnaeus, daily until ecdysis occurred. After ecdysis, spiders were again examined under CO_2 and each larva's presence and position were recorded. Spiders and mantispids recover quickly from CO_2 anesthesia and we have observed no apparent behavioral abberations in either organism. Approximately 2 weeks after reaching maturity spiders were paired in small plastic cages measuring 8.5 × 12.5 × 6.0 cm. Pairings involved not only spiders carrying a larva, but also other spiders that had not been exposed to mantispids. Three types of pairing were made: (1) females carrying a larva and males without a larva; (2) pairings with each sex carrying a larva; (3) females without a larva and males with a larva. Courtship and copulation were carefully observed under a dissecting microscope at 10× when this could be done without disturbing the spiders. After mating or—in some instances—cannibalism, both spiders were examined under CO_2 anesthesia to ascertain any changes in larval position.

Some spiders proved to be subadults and molted only once; other spiders molted twice before reaching maturity. Larval movements on the latter are summarized in Table 19. Movements

Table 19. Movement of *Mantispa uhleri* larvae on immature *Phidippus audax* spiders molting twice to maturity (Exper. 14)

| | Number of Spiders | |
Movement of mantispid larva	♀	♂
Body →[a] pedicel → pedicel	5	2
Body → book lung → book lung	1	7
Body → book lung → pedicel	2	0
Body → book lung → gone	8	0
Total	16	9

[a] Arrow indicates a molt.

of larvae on *P. audax* molting only once are found in Table 20 which additionally condenses and compares all data from Experiments 14 and 16. A total of 21 egg sacs were obtained from the spiders matured and mated in Experiment 14. The larval movements associated with the production of these sacs are summarized in Table 21.

Experiment 15: Larval behavior on third and fourth instar *Phidippus audax*

P. audax were laboratory-reared to third and fourth instars from eggs obtained from laboratory-matured and mated females. After hatching, spiderlings were kept in 7-dram shell vials covered with a piece of nylon screening and fed *Drosophila melanogaster* daily until they reached the desired stage. Boarding of each spider by a single first instar larva of *Mantispa uhleri* was induced in a 2-dram shell vial as in Experiment 14. Larvae were easily visible on these small spiders without CO_2 anesthesia. After boarding by a larva, spiderlings were returned to their original 7-dram vials and fed *Drosophila melanogaster* daily. After each ecdysis, spiderlings were examined under CO_2 anesthesia and any changes in larval position noted.

Most spiders were reared successfully through only two additional molts after larvae had boarded. Significantly more larvae (Table 22) entered the book lungs of the more mature spiderlings.

Experiment 16: Larval Behavior on *Lycosa rabida*

Several adult females of *L. rabida* carrying egg sacs or spiderlings were collected in early September at DSAC. These were kept in individual plastic cages (8.5 × 12.5 × 6.0 cm) until the second instar spiderlings, which remain on their mother's abdomen through this instar, began leaving of their own volition. These spiderlings were isolated singly in the plastic cages just described, modified as follows (Figs. 10 and 11): A 1-cm diameter hole was drilled in the upper right corner in each of the two cage ends and closed with a cork. This permitted food to be added with the plastic top in place. A 1.7-cm diameter hole was drilled in the lower right corner of one side and through this a 2-dram shell vial, filled with water and stoppered with cotton, was

Table 20. Movement of *Mantispa uhleri* larvae on *Phidippus audax* and *Lycosa rabida* spiders (Exper. 14 and 16)

		Number of larvae												
		Position after first molt			Position just prior to last molt				Position after last molt				Survival from book lung after last molt	
Spider species	Sex	Book lung	Ped- icel	Gone	Dead/ dying previ- ously	Book lung	Ped- icel	Gone	Book lung	Ped- icel	Gone	Dead	On Spider	Gone
P. audax molting once to maturity (Exper. 14)	♂	21[a]	3[a]	1	-	-	-	-	-	-	-	-	-	-
	♀	4[a,b]	19[a,b]	4	-	-	-	-	-	-	-	-	-	-
P. audax molting twice to maturity (Exper. 14)	♂	7[b]	2	0	-	-	-	-	7	2	0	0	7[f]	0[f]
	♀	11[b]	5[b]	0	-	-	-	-	1	7	8	0	3[f]	8[f]
L. rabida (Exper. 16)	♂	30[c]	2[c]	0	3	29[d]	0[d]	-	27	2	0	0	29[e]	0[e]
	♀	22[c]	1[c]	0	0	21[d]	2[d]	-	11	2	9	1	12[e]	8[e]

[a] Frequency of larvae entering book lungs and positioned on pedicel significantly different between male and female spiders ($X^2_{adj} = 20.46$, $P < 0.001$, df = 1).

[b] Frequency of larvae entering book lungs and positioned on pedicel significantly different between two groups of female spiders ($X^2_{adj} = 8.458$, $0.005 > P > 0.001$, df = 1).

[c] Frequency of larvae entering book lungs and positioned on pedicel not significantly different between male and female spiders (P via Fischer Exact Test = 1, two-tailed).

[d] Frequency of larvae positioned in book lungs and on pedicel not significantly different between male and female spiders (P via Fischer Exact Test = 0.191, two-tailed).

[e] Frequency of larvae on spiders after last molt and disappearing significantly different between male and female spiders (P via Fischer Exact Test = 0.0003, two-tailed).

[f] Frequency of larvae on spiders after last molt and disappearing significantly different between male and female spiders ($0.005 > P > 0.002$ via Fischer Exact Test, two-tailed).

Table 21. *Mantispa uhleri* larva penetration of *Phidippus audax* egg sacs (Exper. 14)

Position of larva prior to oviposition by spider	Number of spiders[a] spinning egg sacs	Number of larvae after egg sac production		
		In sac	Dead	Missing
On pedicel	17	10	2	5
Elsewhere	1	0	1	0
In book lung	3	2	1	0

[a] One mantispid larva per spider.

Table 22. Movement of *Mantispa uhleri* larvae associated with the first molt on immature *Phidippus audax* spiderlings (Exper. 15)

	Number of spiderlings	
Larval movement associated with molt	Group I: Spiderlings molting from 3rd to 4th instar	Group II: Spiderlings molting from 4th to 5th instar
Body → [a] pedicel	17[b]	13[b]
Body → book lung	8[b]	43[b]
Body → gone	1	10
Total	26	66

[a] Arrow indicates a molt.
[b] Numbers of larvae traveling to the pedicel and to book lungs significantly different in the two groups ($X^2_{adj} = 13.005$, $P < 0.005$, $df = 1$).

inserted to provide spiderlings with constant access to free water. As water evaporated, the cotton pledget was drawn into the vial and continued to provide a wet surface. A 2.7-cm diameter hole on the opposite side of the cage, screened with nylon mesh, provided ventilation. In the bottom of each cage was a small 4-dram vial containing Carolina Biological Supply Instant Drosophila Medium®. A culture of *Drosophila* was thus established in each cage to provide a constant source of food for the second instar spiderlings. Third instar spiderlings received the same food as the preceding stage. *Drosophila* medium was removed from the cage at the beginning of the spiders' fourth stadium and first instar soybean loopers, *Pseudoplusia includens* (Walker), and first instar crickets *Acheta domestica* Linnaeus, were substituted as food. Fifth and sixth instar spiderlings were fed in the same way but with progressively larger loopers and crickets, while sixth instar spiderlings were also given house flies, *Musca domestica*, if accepted. The horizontal water vial was removed when spiders reached the sixth instar and were too large to crawl inside it; the opening thus created was stoppered with a cork. Water was provided for the larger spiders by means of a moistened cotton pledget on a small plastic disc. Beginning with the seventh instar, only house flies were offered as food, and sugar cubes were added to each cage to prolong the life of uneaten flies. This last arrangement remained unchanged until the experiment was complete. As far as possible, spiderlings were reared to the adult stage and were transferred to a clean cage after each ecdysis.

First instar *Mantispa uhleri* were allowed to board the spiderlings when these had reached the fifth or sixth instar. This was accomplished as detailed in Experiment 14 (p. 55). When boarding had been accomplished, the position of the larva was recorded and the spider returned to its plastic rearing cage. After each subsequent ecdysis, spiders were examined under CO_2 anesthesia and any change in larval position recorded. A group of control spiders, never exposed to a mantispid and treated identically, including examination under CO_2 anesthesia, were simultaneously reared.

Approximately 2 weeks after maturing, spiders were paired in a shallow enameled pan (40 × 75 cm) and observations of the ensuing mating or, in some instances, cannibalism, were made. Each male or female spider carrying a larva was paired with a laboratory-reared spider of the opposite sex from the control group. Both spiders were examined under CO_2 for any larval movement after mating. In instances of cannibalism, the remaining spider was examined after it had consumed its partner. Some males carrying larvae were mated more than once, and several were paired with previously mated females to induce cannibalism by the female.

A total of 56 spiders were successfully matured out of 61 originally boarded by larvae. Larval movements noted on these spiders after each ecdysis are summarized in Table 23, while Figure 12 indicates the ultimate fate of these larvae. In the control group, 40 of 44 spiders were successfully matured.

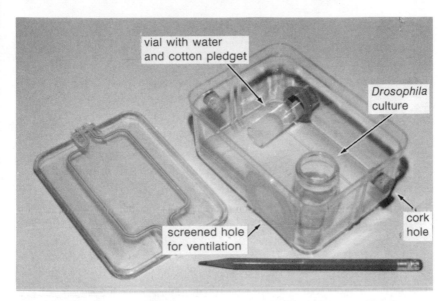

Fig. 10. Rearing cage for *Lycosa rabida* through the fifth instar.

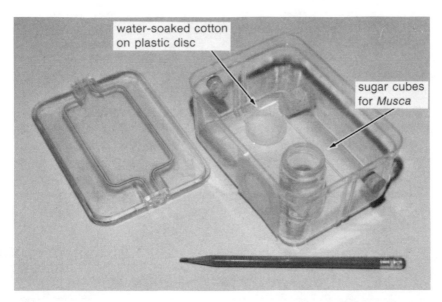

Fig. 11. Rearing cage for *Lycosa rabida* for sixth and subsequent instars.

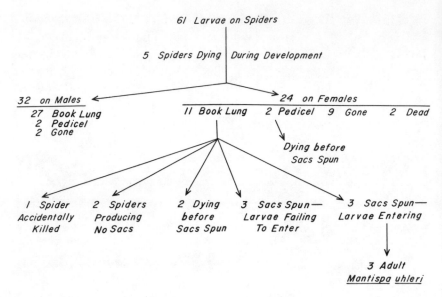

Fig. 12. Fate of *Mantispa uhleri* larvae on developing *Lycosa rabida*.

Table 23. Movement of *Mantispa uhleri* larvae on *Lycosa rabida* spiders reared from fifth or sixth instar to maturity

	Number of spiders	
Mantispid larva movements through several spider molts [a]	♂	♀
Pedicel→ book lung →[b] book lung→book lung	27	10
Pedicel→ book lung→[b] book lung → gone	0	8
Pedicel→ book lung→book lung→ gone→ gone	2	0
Pedicel→ pedicel→ gone→ gone→ gone	1	0
Pedicel→ book lung→ book lung→ pedicel	1	1
Pedicel→ pedicel → book lung→ pedicel	1	1
Pedicel→ book lung→ pedicel→ book lung	0	1
Pedicel→ book lung→ pedicel→ gone	0	1
Pedicel→ book lung→ dead	0	1
Pedicel→ dead→ [b] dead	0	1
Totals	32	24

[a] Arrow signifies a spider molt. Spiders were generally mature at the ninth or tenth instar, so that larvae experienced three to five spider molts. Terminal spider instar was always an adult.
[b] These arrows equal one to four spider molts. All other arrows equal only one molt.

Tables 20, 24, and 25 summarize data from all or some combination of the three experiments. The sites occupied by larvae after first boarding spiders in all three experiments are specified in Table 24. Table 20 catalogues larval movements during spider development in Experiments 14 and 16. A total of 32 male *Phidippus audax* and 34 male *Lycosa rabida* carrying larvae were paired with females. Larval movements during the

Table 24. Initial locations of *Mantispa uhleri* larvae after boarding *Phidippus audax* and *Lycosa rabida* spiders

Location of M. *uhleri* larva on spider	Number of larvae boarding spiders		
	3rd and 4th instar *P. audax*	5th and greater instar *P. audax*	5th and 6th instar *L. rabida*
Pedicel	110	84	54
Under edge of carapace	8	13	5
Between or around coxae	7	4	2
Between sternum and leg bases	2	10	0
Spinnerets	0	12	0
Legs	0	8	0
Totals	127	131	61

Table 25. Transfer of *Mantispa uhleri* larvae from male to female spiders during mating and cannibalism

Spider species	Number of mantispid larvae			
	At spider mating		During cannibalism of ♂ spider by ♀	
	Transfer	No transfer	Transfer	No transfer
Phidippus audax	0	20 in book lung 4 on pedicel	5 from book lung 1 from pedicel	1 in book lung 1 on pedicel
Total		24	6	2
Lycosa rabida	0	20 in book lung	2 from book lung	12 in book lung

mating and cannibalism associated with these pairings are summarized in Table 25.

DISCUSSION

As noted previously, the three experiments reported here have certain features in common and their separate data have a strong interrelationship. Accordingly, we shall discuss our results under four general headings, rather than deal with each experiment separately.

Positions First Adopted by Larvae after Boarding

It is clear from all of the experiments (Tables 19, 20, 22, and 23) that mantisipid larvae are capable of negotiating a spider molt. Since the phenomenon has never been reported before, we were surprised to find that in many instances it is associated with entrance into one of the spider's book lungs. This is, however, never the first location after boarding. The pedicel is the usual site (Table 23), and the larva positions itself anywhere around the exposed circumference of this structure (Fig. 13). Furthermore, on particularly large spiders, a larva may be found in one of the two "pits" formed laterodorsally by the telescoping of the pedicel into the base of the abdomen. In such cases the larva may be completely hidden from view until the abdomen is pulled back to reveal these pockets.

We have observed spiders pick larvae off their bodies during grooming movements of their legs, after which they eat the larvae. In contrast, the pedicel as the resting site for a larva has the advantage of being relatively inaccessible to the spider. Such a location also provides a larva with easy access to the next instar at ecdysis because the spider's old exoskeleton splits laterally along the pedicel to expose the new cuticle only a short distance away. Finally, the pedicel is membranous, as are all other areas that a larva might immediately occupy after boarding. Since they often feed on the spider's blood prior to the production of an egg sac (Redborg and MacLeod, 1983b), larvae must find tissue thin enough for their mouthparts to penetrate.

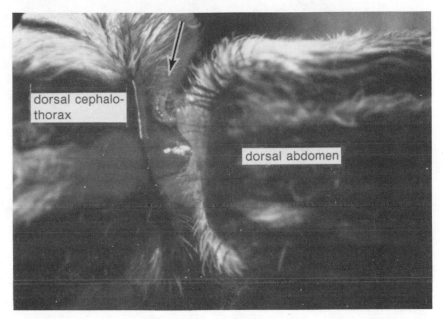

Fig. 13. First instar larva of *Mantispa uhleri* on pedicel of *Lycosa rabida.* Arrow indicates position of larva.

Entry of Larvae into Book Lungs

Movement of larvae into the book lungs of spiders was first discovered when we studied larvae on *Phidippus audax* spiders that were undergoing their final molt to the adult (Experiment 14). The findings (Table 20) revealed a striking difference in larval movements during molting, with respect to the two spider sexes; mantispid larvae entered the book lung of significantly more males than females. We were first inclined to explain this seemingly preferential behavior on the basis of the close proximity of the male's genital opening to the book lungs and the possibility that in such a location a larva might subsequently migrate to the palp at the time of sperm charging and be transferred to the female's abdomen during copulation.

This expectation was not borne out in our analysis of a more general survey of larval movements on immature spiders during two or more molts (Tables 19 and 23). After boarding immature females of *P. audax* (which needed two molts to reach maturity), larvae entered the book lungs at the first molt in significantly larger numbers (Table 20) than did those larvae that had boarded subadult females and experienced only one molt. This demonstrated that larvae did enter the book lungs of female spiders prior to sexual maturity and suggested that larvae made no distinction between males and immature females, responding differently only to adult females.

To study the book-lung-entering behavior further, we had intended to rear the third and fourth instar *P. audax* of Experiment 15 to the adult stage and to follow the movements of the larvae. Unfortunately, nutritional difficulties prevented most of the spiders from passing through more than one or two molts, and entrance into the book lungs could not be studied further until these difficulties were overcome (Experiment 16). In this experiment, virtually all mantispid larvae on both sexes of immature *Lycosa rabida* spiders entered a book lung at the first spider molt (Table 23); frequencies of book lung entry were not significantly different for larvae on pre-adult male and female spiders (Table 20).

Notwithstanding our failure to rear them to the adult stage, the spiders of Experiment 15 did yield some information concerning the relationship of the size of the spider to the time that book-lung-entering behavior is first observed. At the first molt experienced by the larvae, significantly more of them entered the book lungs of spiders molting from the fourth to the fifth instar (Table 22—Group II), than of spiders molting from the third to the fourth instar (Group I). But several of the Group I larvae that were ultimately counted as having gone to the pedicel were seen attempting to enter a book lung immediately following the spider's ecdysis. In both groups of spiders many of the larvae that did enter a book lung had legs or abdomens protruding from the lung slit, suggesting that the book lungs of some of these spiders were simply too small to accommodate a larva. The increased number of book lung entries observed during the ecdysis from the fourth to fifth instar is likely due to the increase in the size of the book lung after this molt.

In all three of our experiments, larvae entered a book lung only in association with a spider molt. The reasons for this are unknown, but conceivably attempts by a larva to enter a book lung during an intermolt period may alert the spider, giving it an opportunity to dislodge its would-be parasite. However, entry is probably comparatively easy immediately following ecdysis while the spider's movements are restricted during the tanning of its new cuticle. It is also possible that the sclerotized margins of the book lung cannot be pried apart as successfully as they can immediately after ecdysis.

Upon first entering the book lung of *L. rabida* (Fig. 14), larvae entered the right and left book lungs with equal frequency (26 left, 31 right, $\chi^2 = 0.439$, $0.9 > P > 0.5$, df $= 1$). Larvae generally remained through subsequent molts in the book lung originally entered. Thus, it is likely that a larva located in a book lung can remain there during a spider's molt, since, if it did leave, there is only a 50% probability that it would re-enter the same book lung. Of 117 instances in which larvae remained in book lungs, 106 occupied the book lung in which they had been located just prior to ecdysis. This result is significantly different ($\chi^2_{adj} = 75.521$, $P < 0.001$, df $= 1$) from the frequency distribution expected if larvae were leaving their respective book lungs and re-entering either the right or left book lungs at random. The few instances where larvae switched book lungs were usually

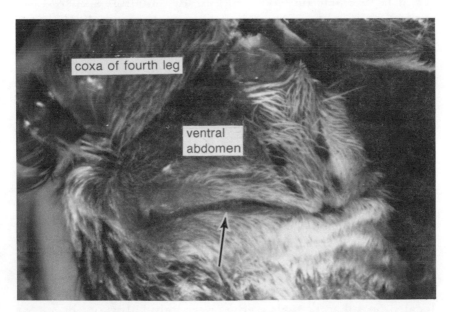

Fig. 14. First instar larva of *Mantispa uhleri* in book lung of *Lycosa rabida*. Arrow indicates position of larva which is barely visible.

correlated with observations of damage (swelling, discoloration, congealed blood) to the book lung previously occupied. These larvae may have been forced to leave the resident book lung during ecdysis because of this damage, or they may have deliberately moved to the more habitable book lung.

Compared to a location on the pedicel of the host spider, a book lung would seem to offer several advantages to a larva awaiting the spider's sexual maturity. Since gas exchange occurs across the book lung membrane, its cuticle should be thinner and thus more easily penetrated by the jaws of a feeding larva. Also, once inside the book lung, the larva runs no risk of being dislodged or damaged by the spider's potentially abrasive grooming movements. Finally, the elevated humidity probably associated with the book lung may benefit the larva.

Larval Movements Associated with the Adult Molt and with Egg-Sac Entry

Our data have suggested that, without regard to the spider's sex, a larva enters the book lung of an immature spider at the earliest molt which produces a sufficiently large book lung. Thus, just prior to their final ecdysis, 50 of the 52 surviving *L. rabida* of Experiment 16 contained larvae in their book lungs, with no significant difference in this regard between male and female spiders (Table 20). Most of these larvae had successfully negotiated two, or more, spider molts in the book lung and there had been only a very low incidence (2 out of 81) of either death or disapperance.

We were, therefore, puzzled to observe a much higher rate of disappearance of larvae associated with the final molt of female spiders. At this time only 12 of the 21 larvae located in a book lung immediately prior to ecdysis were to be found on the spider after the molt (10 of these remained in a book lung, while 2 successfully moved out onto the pedicel—see Table 23). One of the 9 larvae that had previously resided in the book lung died in the process of this molt, but the remaining 8 could no longer be found on their spiders. In contrast, all 29 larvae that had been located on subadult male spiders were found alive on the spider after the molt, 27 of these remaining in a book lung. The number of book lung larvae that succeeded in remaining on their spiders at the final molt varied significantly between male and female spiders (Table 20). This differential behavior with respect to the two spider sexes at the final molt is not restricted to *L. rabida*. A comparison of the movements of larvae on males and females of *Phidippus audax*, which molted twice to maturity in Experiment 14 (Table 20, lines 3 and 4), showed notable differences also. In this case all 7 larvae aboard males survived, compared to only 3 out of 11 on females.

Some light is shed on the fate of the larvae missing from the adult female spiders by the coincidence that one of the spiders carrying a larva (Experiment 16) had been confined in a vial while molting. Within an hour after ecdysis we observed the larva crawling around the bottom of the vial, suggesting that for some reason this larva, and presumably the others, had not been able to remain on their spider through the final ecdysis.

It seems unlikely that simply leaving a female spider at this time is part of a larva's normal developmental behavior since, under natural conditions, the larva—once having left the spider— would have little chance of reboarding because the spider will shortly move away. Additional factors are undoubtedly at work in this process. The possibility that adult spiders actively try to rid themselves of larvae, as well as the possibility that larvae may inefficiently shift from the book lungs to some other location more favorable for entry into an egg sac are considerations to be explored.

In view of the damage to a book lung caused by a resident mantispid larva, as well as the lowered fitness of female spiders whose egg sacs are successfully entered by these same larvae, it would not be surprising to find that spiders have evolved mechanisms for ridding themselves of mantispid larvae. It is possible, therefore, that a larva attempting to remain in a book lung that has optimal access to an egg sac is effectively foiled by some structure or behavior of female spiders.

For several reasons, this does not seem likely to us. That it is not inherently difficult for a larva to negotiate a pre-adult spider molt in this location is demonstrated by our observation (e.g., in Experiment 16) that during 79 of 81 spider molts from one immature instar to another, larvae in book lungs successfully remained *in situ*. It is only in the final molt of female spiders that such a purging mechanism seems to operate, if indeed it exists at all: yet such riddance should also be of advantage to a spider at earlier stages when larvae are feeding as ectoparasites and causing damage to the book lung's delicate structure. Such a mechanism should also be of advantage to male spiders, yet the loss of larvae is restricted to females. Further, since similar results were obtained with two species of spiders belonging to different superfamilies, it seems unlikely that both *P. audax* and *Lycosa rabida* would have independently evolved a similar anti-mantispid mechanism. Finally, the notion that larvae seek to remain in the book lungs of female spiders because this is the most advantageous location from which to enter egg sacs is inconsistent with our findings. With *Phidippus audax* spiders that were undergoing a single molt to sexual maturity, boarding larvae had the opportunity to enter female book lungs but failed to do so in 19 of 23 cases (Experiment 14 and Table 20, lines 1 and 2).

The second possibility seems more reasonable, although it does not account for all of our results. We suggest that larvae leave the book lungs in order to station themselves on an exposed surface (probably the spider's pedicel) so that they can more easily enter egg sacs; we suggest that a gain in the facility with which larvae might enter egg sacs from the spider's pedicel compared to a position in a book lung more than offsets the ineptness with which the book lungs are abandoned. There are four maneuvers that a mantispid larva may perform during the spider's ecdysis: the larva may (1) remain on the pedicel (or sometimes move from another membranous area to the pedicel); (2) move from the pedicel (or elsewhere) into a book lung; (3) remain in a book lung; (4) move from a book lung to the pedicel. From what we have seen of these four activities, we postulate that leaving a book lung is an intrinsically difficult maneuver.

During a spider's ecdysis, the most serious obstacle to the larva is probably the barrier imposed by the old cuticle, since a larva located on a remote area of the exuvia may, after a molt, take too long to get back onto the spider. Since the tanning spider may remain only briefly in contact with its exuvia, a dislocated larva might possibly lose all contact with the spider. In the first two maneuvers noted above, this danger is minimal, since to reach the cuticle of the next instar a larva need only move into the nearby ecdysial tear that runs laterally down the pedicel. The larva can then remain on the pedicel or enter a book lung while the outer margins of the latter are still soft and untanned. Our data show that such larvae, positioned on the pedicel, survive a spider's ecdysis with a high rate of success. In the third situation, a book lung larva need only push against the outward pull of the thin cuticle withdrawn from the book lung during ecdysis in order to tear through it and remain in the same book lung of the next instar. Again, our data indicate that larvae are also adept in this maneuver.

A larva that attempts to leave a book lung during a molt, however, must remain in the book lung long enough to tear through the old cuticle as it is withdrawn from the lung slit, for if the larva leaves too soon it risks being stranded on the exuvia some distance from the spider's new cuticle and may fail to regain the spider. This may have been the fate of the disappearing larvae of Experiments 14 and 16. But a book lung larva that has successfully pierced the cuticle that is being pulled out of the book lung slit must now leave quickly lest it be trapped in the book lung. Just as in entering, leaving a book lung can possibly be done only while the soft cuticle renders the spider immobile. If the cuticle becomes sclerotized before the larva can leave, it may stay in the book lung rather than risk being eaten by the spider aroused by the larva's attempt to escape.

That the spider pedicel is the objective of larvae exiting the book lungs is suggested by the movements of larvae on *P. audax* (Experiment 14) and by the high success rate of larvae in entering egg sacs from this position. Of the 23 larvae on females of *P. audax* that had reached maturity in one molt—so that the larvae did not have to contend with the postulated difficulties of a book lung location—19 of these positioned themselves on the pedicel rather than entering a book lung (Table 20, line 2). Further, as summarized in Table 21, 10 of 17 larvae stationed on the pedicel of this species successfully entered egg sacs, along with 2 of 3 larvae located in a book lung.

The principal difficulty with this explanation is in understanding why mantispid larvae that have entered the book lungs of subadult females leave this position if their rate of success in remaining on the spider is so low. This is particularly puzzling since the data just cited, as well as those of Figure 14 pertaining to *Lycosa rabida*, show that larvae that have remained in book lungs are also successful in entering egg sacs.

Our sample size is not large enough to compare, quantitatively, the success rate of egg sac entry from these two locations. Possibly, additional studies with a larger number of spiders would prove valid the posited advantages of the pedicel over the book lung, and would demonstrate an improved ability to reach sacs from this site which offsets the high rate of dislodgment from the spider that larvae experience when they exit book lungs. Studies of larval movements during the spider's final instars, using the procedures developed in the experiments described here and with additional species of spiders, are needed in order to observe more closely the details of such behavior.

Transfer of Larvae from One Spider to Another

No *Mantispa uhleri* larva attempted to transfer from male to female during our observations of 44 spider matings of *Phidippus audax* and *Lycosa rabida* (Table 25). In the pairings of *Phidippus audax*, some of the females were carrying a larva; transfer behavior of the larva on the male may thus have been inhibited. But since for all *Lycosa rabida* matings the females were free of larvae, inhibited transfer behavior seems unlikely. Several *L. rabida* males were mated more than once to test the possibility that the first mating may convey some necessary preliminary information to the larva, but all of these results were also negative.

Since our laboratory conditions were conducive to the spider's epigamic behavior and successful copulation, it seems a reasonable inference that they would also be suitable for larval transfer, if this should occur. We thus conclude that, although one might have expected selective pressure for such behavior, *Mantispa uhleri* larvae seldom, if at all, transfer from male to female spiders during mating. Although this conclusion might seem counterintuitive, it emphasizes that the theoretical advantage of some heritable trait, deduced by *a priori* considerations, does not ensure that natural selection will necessarily have produced such a trait.

Although the act of spider egg-laying may be similar in most species, enabling mantispid larvae to enter almost any egg sac, epigamic behavior, because of its role as an isolating mechanism, may vary in detail with different kinds of spiders.

Therefore, the utilization of a number of spider species by *M. uhleri* may be incompatible with the evolution of behavior that allows an efficient larva transfer from male to female spider at mating. We argue this since the ability of a larva to synchronize its movements to the details of a particular spider's mating behavior, permitting a transfer from male to female, would be of no advantage to the offspring of that larva unless the same species of spider were often boarded. Consideration of the very large number of spider species naturally utilized by *M. uhleri* makes this latter circumstance seem doubtful.

Nevertheless, larvae boarding male spiders are not necessarily destined to die without reaching maturity. When certain of the *Phidippus audax* males (Table 25) eventually died, larvae were observed to leave the corpse and, presumably, could resume search activity. In nature such larvae may have an enhanced survival and search time because they had fed on spider blood.

Also of potential importance to larvae located on male spiders is a larva's ability to transfer to a new spider during cannibalism. Several larvae in our experiments thus moved from a male to a female spider (Table 25). This action has the advantage of transferring a larva from any spider to any other regardless of spider sex, state of maturity, or even species. Just such an opportunity was observed by one of us (K. E. R.) under field conditions with the collection of two adult females of *Metaphidippus galathea* (Walckenaer), one being consumed by the other. The prey spider had a larva of *Mantispa uhleri* on its pedicel. Although this encounter was permanently disturbed when the spiders were brought into the laboratory and examined under CO_2 anesthesia, the potential advantage for a transfer of the larva to the predator spider existed. More knowledge of how frequently spiders prey on other spiders under natural conditions is needed for a more accurate evaluation of this behavior as it pertains to *M. uhleri*.

7. Species of Spiders Utilized

Most references briefly note the larval association of mantispines with such general statements as "[they] are parasitic in the egg sacs of ground spiders" (Borror and Delong, 1971), in a persistent echo of the fact that the first published account of an adult emergence recorded that it had developed in the egg sac of a lycosid. Except for occasional notes documenting the rearing of a particular mantispid species from the egg sac of a particular spider, virtually nothing has been published about mantispid-spider associations, including the important details of how wide a range of spider species a given mantispid may utilize.

The fact that mantispid species often occur sympatrically suggests they may in some way be apportioning available prey spiders among themselves. As a result, different mantispid species might use spiders of restricted taxonomic groups or might concentrate on spiders found in certain specific habitats. One obvious way to evaluate these possibilities is to collect the data that link adult mantispids with the egg sacs of field-collected female spiders. A second, quite effective method can be used with those mantispid species whose first instar larvae board spiders. Here, with the assumption that the presence of the larva on the spider indicates an eventual egg sac entry, identification of spider and larva establishes the crucial association.

We have used this latter approach in surveying the range of spider species utilized by *M. uhleri*: in our search to find associated larvae, we studied a large, identified collection of spiders, as well as collections we ourselves made in areas where *M. uhleri* was commonly found.

We advance the hypothesis that *M. uhleri* depends on a number of species of hunting spiders, but rarely if ever boards, or eats the eggs of, web-building spiders. Within the broad designation "hunting spider," however, this mantispid appeared not to be restricted to particular taxa. In pursuit of such evidence, we also collected field data on the frequency of larval infestation of the hunting species *Metacyrba undata* and three nonhunting spiders, *Ariadna bicolor*, and two species of *Agelenopsis*, occurring in the same habitat as *Metacyrba*.

METHODS AND RESULTS

To obtain a relatively unbiased view of the variety of spiders used by larvae of *Mantispa uhleri*, we studied the Illinois spider collection of Southern Illinois University at Carbondale (SIUC). Assembled by Dr. Joseph A. Beatty and his students, the collection is the record of an intensive sampling of the spider fauna of southern Illinois where *M. uhleri* is common, and is intended to be as nearly complete a representation of the species in this area as possible. Where species presumed to occur in the region were lacking, special efforts were made to obtain specimens. The magnitude of this effort is apparent in the fact that more than one hundred species have been added to the state list (Beatty and Nelson, 1979). Most important, no spiders were collected on the supposition that they might bear mantispid larvae.

Under a dissecting microscope, and in a Syracuse dish containing alcohol, the entire surface of each spider was carefully scrutinized, the abdomen being retracted to reveal the pedicel. Both book lungs were gently opened with forceps. When larvae were found, they were removed and mounted on slides. The second author's (E. G. M.'s) unpublished key to first instar larvae was used in identification. Any egg sacs attended by spiders and collected with them were also opened and examined for mantispid larvae.

Additional records were gathered by the first author (K. F. R.) from spiders collected at the University of Illinois Dixon Springs Agricultural Center (DSAC) in Pope County and a few other Illinois locales. Over the summer many spiders were collected during the day by sweeping vegetation and examining foliage; at night, lycosids, and occasionally other spider families, were collected with a head lamp. During the winter months, spiders were gathered from beneath the loose bark of trees and from the leaf litter. Living spiders were examined under CO_2 anesthesia. Larvae were identified either by using the key just mentioned or by rearing them to the adult stage.

Spider specimens totalling 5,761 were examined from the SIUC collection, and the species and number of individuals were listed in a sequence of families recommended by J. A. Beatty. Sixteen larvae of *M. uhleri* were found, all of which were associated with hunting spiders of the Lycosoidea and Clubionoidea (see Appendix I [p. 101] and Table 26).

Appendix II [p. 117] summarizes fundamental data for all species of spiders which we have found to be utilized by *M. uhleri*, and includes larvae from the SIUC collection (extracted from Table 26) as well as records of 110 additional larvae which we obtained from spiders collected at DSAC and other locations. The arrangement is alphabetical by family, genus, and species. For each species is listed information on all larvae removed from spiders of that species including the sex and maturity of the spider, location of larva on spider, the site where collected, and the date when acquired. No locale is specified if the spider was part of our supplemental collecting at DSAC. If more than one larva was found on a single spider this fact is noted. The total larval associations involved 31 species of spiders distributed in 22 genera (Table 27), and represents all but two of the families of the superfamilies of hunting spiders noted above.

During the winters of 1974-75 and 1975-76 a survey of the occurrence of larval *M. uhleri* on the spiders *Metacyrba undata* and *Ariadna bicolor* was conducted. These spiders were collected from silk retreats beneath the bark of shagbark hickory, *Carya ovata* K. Koch, from ground level to a height of approximately 10 feet. The woods surrounding DSAC were entered from several arbitrarily chosen points and all shagbark hickories encountered were examined for loose bark. All spiders that could be located on each suitable tree were collected and brought into the laboratory where, under CO_2 anesthesia, they were examined for larvae.

In 1976 a similar survey was made of the number of larvae on two agelenids, *Agelenopsis kastoni* Chamberlin and Ivie and *A.*

Table 26. Summary of *Mantispa uhleri* Banks-spider associations from the examination of the SIUC Collection

Spider suborder, superfamily, family	Spider genera		Spider species		Total spiders	
	Examined	Larvae present	Examined	Larvae present	Examined	Larvae present
MYGALOMORPHAE						
ATYPOIDEA						
Antrodiaetidae	2	0	2	0	68	0
CTENIZOIDEA						
Ctenizidae	1	0	1	0	4	0
ARANEOMORPHAE						
DICTYNOIDEA						
Amaurobiidae	2	0	3	0	12	0
Dictynidae	2	0	9	0	218	0
Oecobiidae	1	0	1	0	2	0
Uloboridae	1	0	1	0	10	0
DYSDEROIDEA						
Dysderidae	2	0	2	0	7	0
SCYTODOIDEA						
Scytodidae	2	0	2	0	48	0
ARANEOIDEA						
Pholcidae	2	0	2	0	22	0
Theridiidae	17	0	36	0	810	0
Linyphiidae	26	0	40	0	515	0
Araneidae	23	0	44	0	1,141	0
Symphytognath- idae	2	0	2	0	2	0
Mimetidae	2	0	3	0	18	0
LYCOSOIDEA						
Agelenidae	4	0	11	0	248	0
Hahniidae	2	0	3	0	21	0
Pisauridae	2	1	8	1	76	1
Lycosidae	7	0	24	0	614	0
Oxyopidae	1	0	3	0	289	0
CLUBIONOIDEA						
Gnaphosidae	10	0	21	0	108	0
Clubionidae	7	2	24	2	350	2
Anyphaenidae	3	1	7	1	118	1
Thomisidae	11	3	32	3	406	3
Salticidae	21	4	38	5	654	9

Table 27. Taxonomic summary of all spider species utilized by *Mantispa uhleri*

Spider Superfamily, Family	Spider genera examined		Spider species examined	
	Total	Larvae present	Total	Larvae present
LYCOSOIDEA				
Agelenidae	4	1	11	1
Hahniidae	2	0	3	0
Pisauridae	2	2	8	2
Lycosidae	7	2	24	4
Oxyopidae	1	0	3	0
CLUBIONOIDEA				
Gnaphosidae	10	1	19	1
Clubionidae	7	2	24	2
Anyphaenidae	3	1	7	2
Thomisidae	11	5	32	6
Salticidae	21	8	38	13
Totals	68	22	169	31

emertoni Chamberlin and Ivie. Although all of the adult specimens collected belonged to these two species, many specimens were immature and could not be identified with certainty. Therefore, all specimens are hereafter referred to as *Agelenopsis* spp. These collections were made in April, after the weather had moderated sufficiently to allow the spiders to move from their overwintering sites and construct funnel webs in the leaf litter and around shrubs and fallen branches. The areas searched were generally the same as where *Metacyrba* and *Ariadna* had been collected during the preceding two winters. All funnel webs found were examined and spiders occupying them were coaxed into collecting vials and preserved in 70% ethyl alcohol. These spiders were examined for larvae as described previously for specimens in alcohol.

Incidence of larvae on these spiders is recorded in Table 28; analysis reveals almost no larvae from *Ariadna bicolor* and *Agelenopsis* spp., while 8.5% of *Metacyrba undata* were infested with mantispid larvae.

Table 28. Number of first instar larvae of *Mantispa uhleri* Banks found on overwintering spiders collected at Dixon Springs Agricultural Center

Spider collected	Total specimens examined	Specimens without larvae	Specimens with larvae
Metacyrba undata	484	443	41
Ariadna bicolor	148	148	0
Agelenopsis spp.	216	215	1

DISCUSSION

Very early in the course of collecting the spider specimens it became apparent that, in nature, larvae of *Mantispa uhleri* were boarding a wide variety of hunting spiders. During this period web-building spiders were also collected, but as no mantispid larvae were found on them, collecting unavoidably concentrated on those groups of spiders that harbored larvae. Thus, we realized that to single out effectively the major groups of spiders used by *M. uhleri* it would be necessary to obtain an impartial sampling of spiders. The SIUC collection fulfilled this requirement and permitted us to test the hypothesis that, in nature, searching first instar larvae board primarily, or solely, spiders belonging to the non-web-spinning, hunting groups.

We found 16 larvae of *M. uhleri*, 13 on spiders, 3 within egg sacs (out of a total of 35 sacs) attended by spiders (Table 26 and Appendix I), and, as noted, these were associated solely with the hunting groups (Table 26, superfamilies Lycosoidea and Clubionoidea). A formal statistical analysis of these data was inappropriate since, although gathered at random with respect to the presence of mantispid larvae, the collecting effort was not necessarily standardized with respect to the species sampled or to the time of year. The complete absence of larvae associated with the 2,987 specimens of nonhunting species (Table 26, Families Antrodiactidae through Mimetidae) is impressive and, we strongly feel, consistent with our hypothesis. The extensive additional data on hunting species collected at DSAC (App. II) and the absence of larvae on nonhunting groups collected from that locale support our generalization.

That members of nonhunting groups of spiders may occasionally be utilized by *M. uhleri* is suggested by some of our laboratory observations. For instance, females of even a web builder like *Achaearanea tepidariorum* are readily boarded if confined in a vial with a larva and, by means of a cotton plug, restricted to a small space so that they cannot suspend themselves within a web. Likewise, although none of the 148 *Ariadna bicolor* of the DSAC field sample carried a larva, even though collected from the same microhabitat that yielded abundant larvae-bearing *Metacyrba undata,* in the laboratory this spider species is boarded with ease when restrained from spinning a retreat. The single larva taken from our substantial series of *Agelenopsis* spp. compared to the large number taken from *Metacyrba undata* also suggests that *Mantispa uhleri*'s failure to utilize nonhunting spiders is related more to the infrequency of direct contact than to avoidance of them.

The data of Appendix II also show that, at least insofar as sexually mature spiders are concerned, both males and females are boarded in nature (48 male spiders bearing larvae, 36 females). This corroborates an inference from our earlier experimental findings (Table 16) that mantispid larvae do board male spiders under natural conditions. Many of these larvae were found in the book lungs of both male and female spiders. This is again consistent with our laboratory findings (Table 22)— particularly our contention that entrance into the book lungs is not associated solely with male spiders. In our field sample, however, for a number of the larvae-bearing species, the adults are probably too small for larvae to enter the book lungs. Our findings, related to the preferential residence in the book lungs by mantispid larvae during the development of immature and adult male spiders, are thus appropriate only with the larger spiders such as those we used in our experiments.

Mantispid larvae are generally thought to be rare. Indeed, even after observing numerous larval boardings in the laboratory, we pessimistically felt that field confirmation of this behavior might be like looking for the proverbial needle. Fortunately, we found that larvae were quite common, at least at Dixon Springs (Table 28). One of twelve overwintering *Metacyrba undata* had been boarded by a larva, which suggests that mantispids prey on the eggs of some spider species much more commonly than has been previously supposed.

8. The Mantispid Seasonal Cycle

Our investigation of this predator-prey relationship has demonstrated what is probably a complex seasonal cycle and points to the fact that *Mantispa uhleri* relies on a large number of spider species to govern—by means of their own biological time-clocks—the timing of mantispid seasonal development. A general model for *M. uhleri*'s seasonal cycle can be developed and related to results of two years of seasonal distribution records of larvae and adults derived from collections made in southern Illinois. Supplementary information collected in other Illinois locales has also been incorporated.

METHODS AND RESULTS

Adult *M. uhleri* were collected in two light traps, which were similar to one illustrated by Metcalf and Luckmann (1975, p. 329) and were set up about 10 yards from the edge of extensive woodlands in the Shawnee National Forest at DSAC in southern Illinois. The traps were situated about 50 yards apart and each trap consisted of four 15-watt ultraviolet fluorescent tubes placed above a wide-mouthed metal funnel leading into a screened cage approximately 2 feet on each side. A wooden roof above the lights prevented rain from entering the traps. Metal flanges on both sides of the lights aided in the capture of flying insects. The distance from the lights to the ground was approximately 10 feet. In theory, insects attracted by ultraviolet radiation hit the lights or flanges and fell through the funnel into the box below. Frequently, dry ice, insulated and covered by crumpled newspapers, was placed in the bottom of a waste basket-sized container immediately below the funnel, so that insects passing through the funnel were anesthetized by the sublimating carbon dioxide. This reduced mutilation of mantispids by crawling beetles and fluttering moths and left most of them in an apparently undamaged condition when released the following morning.

With the exceptions noted later and in Figures 15 and 16, the traps were operated every night from mid May 1974 to mid April 1976. Traps were checked in the early morning and the number and sex of any *M. uhleri* were recorded. Except for adults that were kept for rearing purposes, undamaged mantispids were marked and released next to the traps the morning after their capture, in the hope that recapture would give some indication of their longevity under natural conditions. Small spots of colored fingernail polish were placed on the undersurfaces of the wings in a characteristic pattern so that individual insects could be identified if re-collected. Temperature data were recorded for the periods of collection by means of a hygrothermograph located near the traps.

Figures 15 and 16 record the number of *M. uhleri* collected during 1974 and 1975, respectively. High and low temperatures for each collecting night are indicated in both figures. The highest temperature usually occurred at sunset and the lowest at sunrise. Approximately equal numbers of male and female mantispids were collected; there was no indication of a male-biased sex ratio such as that observed by New and Haddow (1973) from their light-trap collections of mantispids at Entebbe, Uganda, in 1961 and 1962. In order to evaluate the severity of the winters preceding our collections, the average weekly temperatures were plotted for December-April of 1973-74 and 1974-75 (Fig. 17).

During the summer of 1974, eggs were obtained on 6 July from the earliest trap-collected female *M. uhleri* that produced fertile eggs in captivity. Larvae from these eggs were reared in an unheated screened insectary at DSAC so that the generation time under field conditions could be estimated and an indication obtained as to when to expect the earliest possible emergence of adults of a second summer generation. The first adults from these eggs eclosed on 12 August (Fig. 15), representing a field-

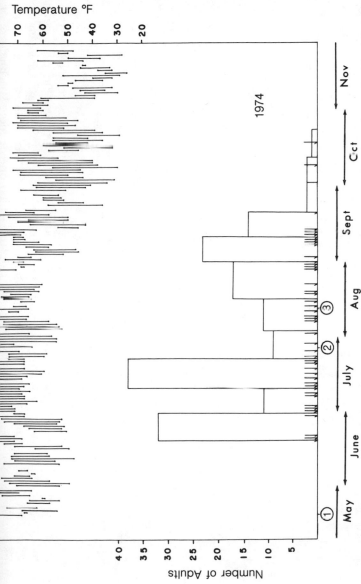

Fig. 15. Adult *Mantispa uhleri* collections, 1974. Each bar represents the number of mantispids collected during 10 nights of actual trap operation. Nights on which the traps were not run are not included; some of the bars are consequently wider than others. Above the histogram are given the high and low temperatures for each collecting night, the extremes being connected by a line. Arrows at the base of the histogram signify nights during which one or more *M. uhleri* specimens were collected.

① first operation of traps; ② earliest emergence of a second generation, assuming a generation time of 37 days from the first appearance of adults in the traps; ③ first adults reared in outdoor insectary from eggs of first ovipositing female.

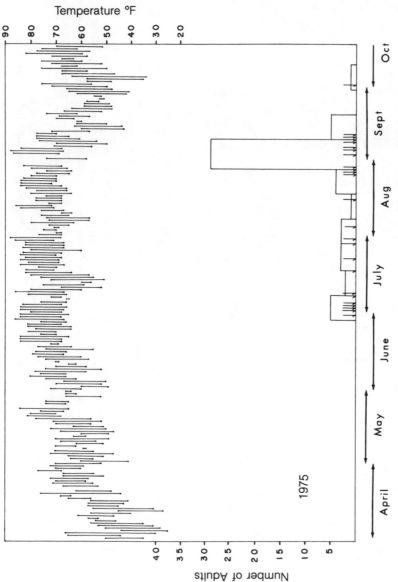

Fig. 16. Adult *Mantispa uhleri* collections, 1975. Each bar represents the number of mantispids collected during 10 nights of actual trap operation. Nights on which the traps were not run are not included; some of the bars are consequently wider than others. Above the histogram are given the high and low temperatures for each collecting night, the extremes being connected by a line. Arrows at the base of the histogram signify nights during which one or more specimens of *M.*

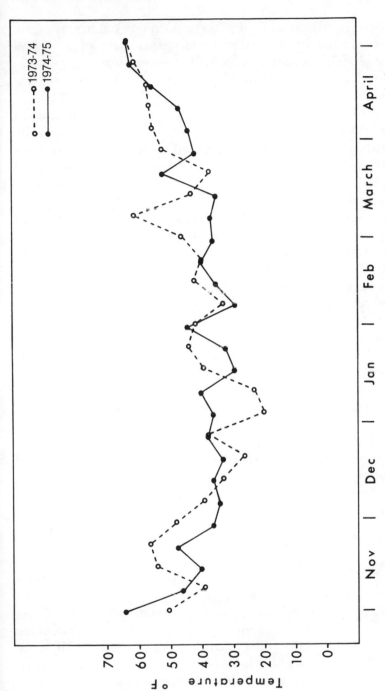

Fig. 17. Winter temperatures preceding summer collections. Mean temperatures for 7-day periods were calculated by averaging four daily temperatures, recorded at 6-hour intervals beginning at 12 noon. Each point represents the midpoint of one 7-day period.

laboratory generation time of 37 days. Also marked in this figure is an estimation of the earliest emergence date of a second summer generation, assuming a 37-day generation time and that eggs were laid in the field on 20 June, the first day an adult was collected.

A total of 88 adults of *M. uhleri* were marked and released in 1974 and 34 in 1975. Three were recaptured in 1974: one male was recaptured twice—on 10 September and 12 September—having originally been collected 7-9 September; a second male caught 7-12 September was recaptured 13 September. No recaptures occurred in 1975.

Several adult females of *M. uhleri* captured toward the end of the 1975 season (September and October) were kept in the insectary under natural conditions to observe any indications of an adult diapause: none was noted. Females produced eggs until they eventually succumbed to the cold weather. Successive egg clutches developed and hatched until, they too, were halted by the cold. No fertile eggs that failed to hatch in the fall survived the winter. The last eggs to hatch were laid on 1 October and hatched 27 October; the confined, unfed first instar larvae displayed no signs of arrested development. They exhibited typical searching behavior as previously observed in the laboratory at long photoperiods. In approximately one week they all died.

On the basis of our trapping data, we attempted to find *M. uhleri* in naturally occurring egg sacs of spiders that had overwintered as subadults or adults. Our survey was conducted 10-27 June 1979 during four collecting trips to wooded areas at Lake of the Woods, Champaign Co., Illinois, and along the banks of the Vermilion River, Vermilion Co., Illinois. We looked for egg sacs under the loose bark of shagbark hickory. The mantispids collected were brought into the laboratory and kept at 25°C and a photoperiod of L:D = 16:8. The dates of their adult emergence were recorded (Table 29).

A few egg sacs of *Phidippus audax* and *Metacyrba undata* were found, but by far the majority of the egg sacs discovered belonged to the crab spider, *Philodromus vulgaris*. All seven mantispid associations (Table 29)—five of them verified *Mantispa uhleri*—were with egg sacs of *Philodromus vulgaris*. In six of the

Table 29. Presence of *Mantispa uhleri* in *Philodromus vulgaris* egg sacs collected in 1979

Date of collection in 1979	Number of egg sacs examined	Number and developmental state of mantispids[a] found	Date of adult *M. uhleri* emergence
10 June	5	1 third instar	23 June
17 June	10	1 prepupa in cocoon	killed during collection
19 June	not recorded	1 pupa	29 June
27 June	not recorded	2 pupae[b]	2 July collected dead[c]
		1 pharate adult	
		1 empty cocoon[d]	

[a] Each association from a separate egg sac.
[b] Pupae in separate egg sacs. One sac attended by female spider which had spun a second sac.
[c] Specimen recently dead: found next to the egg sac from which it had exited.
[d] Adult recently emerged: the attending female spider had spun a second egg sac.

associations, the egg sacs were guarded by the female spiders that had constructed them. The one unattended egg sac containing a mantispid could still be categorized as recently made because the sac contained a few live spiderlings that had not been eaten earlier as eggs by the mantispid.

We were amazed at the number of ragged and tattered mantispid cocoons that could still be detected in the remains of spider egg sacs from years past. Thirty to forty such cocoons were found. We wonder how many naturalists have observed but not recognized these relics. Figure 18 illustrates the typical appearance of cocoons of *Mantispa uhleri* as we found them in *Philodromus vulgaris* egg sacs.

Seasonal data concerning the occurrence of first instar larvae of *M. uhleri* on wild-caught spiders in southern Illinois (Appendix II) are summarized in Table 30. A total of 123 larvae were recovered from 115 spiders, some of the boarded spiders bearing two, or even three, larvae. Our procedure was not standardized in making these spider collections; therefore, no conclusions as to the relative abundance of these larvae can be drawn. It is important, however, to note their occurrence in nearly all months of the year.

Fig. 18. Naturally occurring cocoons of *Mantispa uhleri* in the egg sacs of *Philodromus vulgaris* collected beneath the bark of shagbark hickory trees. Arrow indicates the position of a cocoon.

A. Cocoon to the left, recently made (sac opened to reveal cocoon). To the right, old tattered cocoon—from past years.

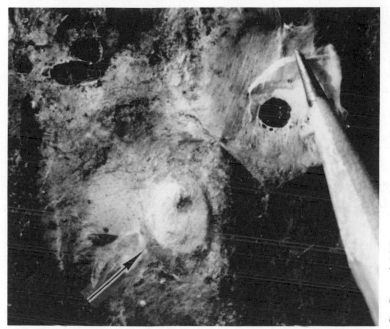

C. Sac of 18B opened to reveal cocoon.

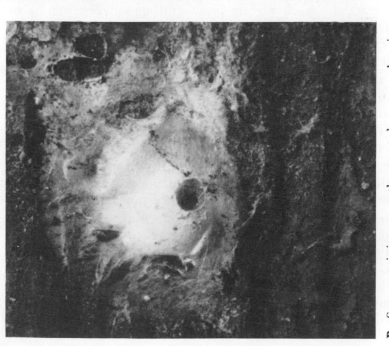

B. Sac containing recently vacated cocoon showing exit hole of the pharate adult.

Table 30. Monthly collections of first instar *Mantispa uhleri* on spiders in southern Illinois

Month	Number of larvae	Month	Number of larvae
January	16	July	1
February	20	August	8
March	16	September	31
April	9	October	11
May	6	November	3
June	2	December	0

DISCUSSION

Mantispa uhleri has, along with some internal parasitoids, the unusual attribute of a larval diapause that is apparently induced and terminated by another, host, organism. Thus, after boarding a spider, a larva ceases development and, in the case that the spider is a pre-adult, waits for the spider to attain sexual maturity and produce an egg sac. The larva then enters the sac and resumes development. The cues for diapause induction and termination are presumably linked to the behavior of the spider (i.e., the act of boarding by the larva, and egg sac construction by the spider, respectively) but, since larvae ingest spider blood, vital chemical cues may also pertain. On larvae reared in pseudosacs at short photophases and/or cool temperatures, we have seen no developmental effects that might indicate the mechanism of any diapause control by these factors of the physical environment. Thus, *M. uhleri* apparently depends on the spider it boards to monitor the crucial environmental cues necessary for the seasonal timing of its reproductive cycle.

The numerous spider species that are boarded by *M. uhleri* spin their egg sacs at varying times of the year. Even if only a fraction of these spiders are successfully used by *M. uhleri* in producing its own adults, one would expect these adults to emerge at several different times during a season. Thus, if, on the day after a mantispid larva has boarded it, a spider spins an egg sac, the larva will begin development within a day. But if the larva has boarded a young spider that is overwintering before maturation, development of the mantispid adult will be resultantly delayed until some characteristic time the following year.

All of our field observations suggest that a spider-inhabiting first instar larva is the only overwintering stage of *M. uhleri*. In both years of our study, adult mantispids did not appear in the traps until late June, the initial dates in these two years differing by only eight days. During May and early June of both years many days were warm enough to lure mantispids to the traps, and if *M. uhleri* had overwintered as a fully fed larva, pupa, or adult, one might expect that adults would be seen before late June. But larvae and pupae of *M. uhleri* collected from the egg sacs of *Philodromus vulgaris* in June 1979 had most certainly not overwintered in that condition, since the egg sacs had been spun only that spring. *P. vulgaris* overwinters as a subadult and matures during April in New England (Kaston, 1948). The specimens of *Mantispa uhleri* that we collected (Table 29) had without doubt overwintered as first instar larvae on the spiders. Also in support of our conclusion is the negative evidence characterized by our repeated failures to collect any other over-wintering stage, our observations in the insectary of fall-collected adults and their offspring, and our inability to bring about laboratory-induced diapause in any other stage.

The first adult mantispids of the year probably emerge from late spring or early summer egg sacs of spiders that had overwintered as adults or subadults. For these spiders, feeding activity would probably begin in March or early April, and egg sacs would be produced sometime thereafter. Assuming a developmental time of 28 days (Table 3), our first adult mantispids (Figs. 15 and 16) probably emerged from egg sacs produced sometime in late May or early June. It is noteworthy that the mantispids that we know entered egg sacs under just such conditions (Table 29), albeit slightly farther north, emerged as adults during the same seasonal period as the earliest adults trapped in southern Illinois.

Larvae will also have overwintered on less mature spiders of other species that will be maturing and spinning their egg sacs later in the year. Thus, it might be expected that emergence of the overwintered generation of *M. uhleri* would occur through-out the summer. We suggest that this emergence curve might be skewed toward the earlier months of the warm season to the extent that the incidence of larvae on species of early egg-sac-spinning spiders may be greater than on those of species spinning later since, assuming comparable population sizes of these different groups, the spiders spinning early would have been available for a longer period during the previous year, thereby increasing the likelihood of their being boarded by a larva.

At some time during the summer the eggs produced by the mantispid adults that have matured from the overwintered generation of larvae should begin to appear. Some of the hatching larvae from these mantispids would board spiders destined to spin egg sacs that same year. As a result, a second generation of adult *M. uhleri* should emerge, possibly overlapping the still-continuing emergence of the overwintered generation.

The complexity of this cycle may be illustrated with *Phidippus audax* and *Lycosa rabida*, from each of which we have taken larvae of *Mantispa uhleri* in the field. Our observations in Illinois agree with those of Kaston (1948), who concluded that both of these spiders have one generation per year in New England. *Phidippus audax* overwinters as a subadult in a silken retreat, matures in the spring and soon afterward spins its egg sacs. The offspring then develop slowly through the summer. *Lycosa rabida* overwinters as a small spiderling—probably in the leaf litter—and sexual maturity and sac production take place in mid to late summer.

We expect that an overwintered larva on *Phidippus audax* would enter an egg sac early in the year and would emerge as an adult, would mate and produce larvae while overwintered larvae on *Lycosa rabida* were still on the spider. These early summer, first instar larvae—the offspring of those overwintering on *Phidippus audax*—might then board small immature *P. audax* spiders that would have hatched a few weeks before, overwinter on subadults of these spiders and enter an egg sac the following spring, approximately one year after hatching from the egg. Some of the same group of early summer larvae might board nearly mature *Lycosa rabida* spiders and enter egg sacs within a few weeks, so that the resultant adults would be emerging during the same summer. This simplified analysis considers only two spiders: the true situation is certainly even more complicated because of the great number of spider species utilized by *Mantispa uhleri*. Peck and Whitcomb (1978), in a study of the phenology of cursorial spiders in the south central United States, demonstrated "a shifting numerical dominance from species to species through the seasons as the adults of one species displaced those of another closely related one." This is precisely what we suggest is occurring with the spiders utilized by *M. uhleri*. In fact, two of the spiders from the Peck and Whitcomb study, *Schizocosa ocreata* Hentz and *Lycosa rabida*, are included in the host range of *Mantispa uhleri* (Appendix II). *Schizocosa ocreata* adults appear in midsummer (June-August), between the early (April-May) maturity of *Phidippus audax* and late (July-September) maturity of *Lycosa rabida*.

The number of adult *M. uhleri* emerging at any particular time of year, then, should be directly proportional to the number of larvae on various spider species spinning egg sacs approximately one month earlier. For spiders spinning in early summer, these totals derive entirely from overwintered larvae and are probably proportional to the length of time the spiders were available for boarding the previous year and the increasing numbers of mantispid larvae present through the previous summer. For later spinning spiders, producing egg sacs after new summer larvae have appeared in the field, the number of larvae on spiders at the time of egg sac production should increase as the season advances and be composed of an ever-decreasing proportion of overwintered larvae and an increasing proportion of larvae that have hatched during that summer.

A given lineage may conceivably have one, two, or even three generations a year, depending on the host spider species. Three generations would seem to be a maximum, based on our calculated 37-day generation time. Three generations should be relatively uncommon, however, since each larva in a given lineage would have to locate spider eggs immediately (by direct sac penetration or boarding a spider about to spin) twice in succession. An average of two generations per year seems more likely.

This analysis predicts a complex emergence curve, compounded of two separate curves representing overwintered and summer generations. The two components of the overall curve do not imply two separate generations of *M. uhleri* each year, but rather the combined effect on the overall emergence curve by two groups of larvae with distinctly different histories. The component of adults of the additional summer generations is probably skewed toward the later months of the year because, with the passage of time, there is the likelihood of an increasing frequency of larvae as a result of the growing population of searching larvae and the greater time with which the spiders are exposed to these new summer larvae. In reality the resultant emergence curve could vary considerably depending on the proportionate contributions from the different species of spiders. Peaks could result from large numbers of certain spider species spinning their sacs at particular times.

If our prediction of multiple, overlapping generations is correct, we would expect to find first instar larvae on spiders throughout the summer. The data in Table 30 are consistent with this expectation. Although the methods used in collecting these larvae-bearing spiders were not standardized, it is our subjective impression that at no time were such spiders particularly difficult to secure.

Data obtained through light-trapping should furnish a relative index of the actual population at the times of collection. Thus, if *M. uhleri* is long-lived in its natural habitat, we would expect to find the capture frequency of adults of the population produced by our predicted emergence patterns to be gradually increasing through much of the season.

It is of interest, then, that our data for 1974 (Fig. 15) show a decrease in adult collections during midsummer, suggesting that this curve is more closely approximating emergence. Although our procedures were not originally designed to collect data directly bearing on the following discussion, we shall briefly touch on several circumstances that might cause data drawn from a standing population to approximate the form of an emergence curve. It is possible, for example, that adults of *M. uhleri* are, in fact, relatively short-lived, so that samples actually are being drawn from such an emerging population. Although we have kept adults alive in the laboratory for several months, it is difficult to equate this with natural conditions where heavy predation or sensitivity to changing weather conditions might produce a much shorter mean life span.

A further possibility is that individuals of *M. uhleri* are attracted to our traps only during a brief period of their life span, such as while searching for mates or during a time of migration or intensive feeding immediately after eclosion. Under these circumstances, too, our collecting data would approximate those derived from an emergence curve. One can visualize additional, special circumstances relating to the stress of capture or the weather conditions. It is obvious that our study is only a preliminary examination of a situation in which many additional data bearing on the basic features of the ecology of the adults of *M. uhleri* are needed.

The collecting data for 1975 (Fig. 16) seem completely different from those of 1974. We believe it likely, however, that the overwintering and additional summer generation curves are independent in some degree. Thus, factors that might affect the overwintering population and lower the number of mantispids collected during the first part of the year might not substantially alter the emergence curve of additional summer generations if the offspring larvae from the overwintered survivors encountered summer spiders often enough. This may have happened in 1975, since the number of mantispids collected from late August to early September was virtually the same in both 1974 and 1975.

We have no idea what factors might have lowered the overwintering population of 1975 or, perhaps, enhanced the winter survivals of 1974. There seemed to be no obvious temperature differences between the winters of 1973-74 and 1974-75 (Fig. 17). Only additional studies can demonstrate whether such year-to-year fluctuations are the rule for *M. uhleri* and from what they derive.

9. Summary

Mantispa uhleri is a member of the neuropteran family Mantis-pidae and develops exclusively in the egg sacs of spiders. In our field and laboratory investigations of this species we have attempted to discover how the larvae of this species locate food, to identify the kinds of spiders utilized, and ultimately to provide a model for the mantispid's seasonal cycle. Culture methods that we devised for successful laboratory maintenance of *M. uhleri* and several other mantispid species should be applicable to additional studies throughout the family. We have recorded measurements and descriptions of the developmental stages of *M. uhleri,* and have described its mating behavior.

Every three to five days, mated adult females of *M. uhleri* each lay a clutch of eggs containing from several hundred to several thousand eggs that are attached to the substrate by a short stalk. The number of eggs per clutch is proportional to the body size of the ovipositing female. Extreme variation in adult size is quite common in this species (and other mantispids, as well), due to the varying amount of egg material in the spider egg sacs which larvae enter. Larvae are "locked in" to their food supply, having no way of locating additional eggs. A third instar larva of *M. uhleri* thus begins spinning a cocoon whenever its supply of spider eggs is exhausted.

After hatching, first instar larvae of *M. uhleri* must find their own food supply of spider eggs. We have no evidence that females oviposit in any preferential areas that would benefit the larvae. Two distinctive tactics are used by mantispid larvae to locate spider eggs: the direct penetration of an egg sac already constructed in the environment, or the boarding of a female spider prior to sac production and entering of the egg sac at the time of spinning. Some mantispid species employ only one of these tactics. *Mantispa viridis*, for instance, penetrates egg sacs only, and displays no interest in spiders. *Mantispa uhleri*, however, facultatively uses both egg sac penetration and spider boarding, although its penetration technique is neither as rapid nor as predictable as that of *M. viridis*. When larvae were given a choice of penetrating an egg sac or leaving its vicinity, they only occasionally penetrated the sac. But when given a choice between a restrained spider or an egg sac, the larvae always boarded the spider. These results suggest that immediate penetration of an egg sac is of minor importance to *M. uhleri* and that the principal larval activity consists of boarding spiders. Although these experiments were not designed to test for the utilization of chemical or tactile cues in locating egg sacs or spiders, no evidence suggested that either are located by anything but random searching by the larvae.

To enter an egg sac a larva appresses its head to the surface of the sac and abrades the silk by back-and-forth head movements. As it enlarges the opening, the larva pushes forward until it has entirely invaded the egg sac and is out of sight beneath the silk. Once it is inside the sac, several factors may interfere with a larva's survival, including age of the sac and simultaneous invasion of the sac by other larvae.

Larvae will board spiders of either sex and any state of maturity, although they apparently fall within the prey size-range of very small spiders, making successful boarding of these spiders difficult. Larval boarding seems not to be affected by the presence of another mantispid already on the spider. After climbing on a spider, larvae of *M. uhleri* preferentially wrap themselves around the spider's pedicel. They negotiate the molts of immature spiders with remarkable success, and at the first molt that produces a sufficiently large spider, they enter its book lungs, regardless of the spider's sex. These larvae generally remain in the book lungs during the spider's subsequent development. At the molt which produces a mature female spider, however, larvae attempt to leave the book lungs for the pedicel, probably in anticipation of the spider's impending oviposition. This maneuver is often unsuccessful; many larvae are dislodged from the spider and cannot regain their position. We did not find that any larvae transfer from mature male spiders to females during mating; such transfers, however, did sometimes occur while male spiders were being cannibalized by females.

During their tenure on a spider, larvae feed on spider hemolymph by penetrating the thin cuticle around the pedicel or within a book lung, but they do not engorge. During egg sac production the larvae enter the egg sac and only then resume development by piercing and draining the contents of the spider's eggs. There are three larval instars. The mature third instar spins a cocoon within the spider egg sac, using silk from its Malpighian tubules. The pharate adult bites its way out of the cocoon and egg sac and crawls some distance away before undergoing the final ecdysis to the adult.

The number of spider species used by larvae of *M. uhleri* was investigated by the examination of a collection of over 5,000 spiders that had been independently assembled during a study of the spiders of southern Illinois. This sample, unbiased with respect to the likely presence of mantispid larvae on spiders, included sixteen first instar larvae of *M. uhleri*. All of them were found on species of the hunting groups Lycosoidea and Clubionoidea. This, together with the absence of *M. uhleri* larvae on the nearly 3,000 specimens of nonhunting spiders, suggests that first instar *M. uhleri* are often or always associated with hunting spiders but rarely, if ever, with web spinners. A large sample of spiders collected by the authors from a restricted locale in southern Illinois supports this conclusion. The two collections together provide association of over one hundred larvae of *M. uhleri* with 31 species of hunting spiders distributed among 21 genera and almost every family of hunters.

Our field studies of *M. uhleri* in southern Illinois showed that this species overwinters as first instar larvae on spiders; it seems likely that this is the only overwintering stage in this area. Beginning in the spring, the larvae enter the egg sacs of the spiders on which they overwintered, the time of entrance being controlled by the phenologies of the different species of spiders. Adults of these larvae begin emerging in late June or early July. In midsummer, probably before all the overwintered larvae have entered sacs, new larvae produced by early emerging adults appear in the field. The timing of the transformation of these new larvae to the adult stage depends also on the species of spiders that they board. Direct penetration of egg sacs at any time can be considered comparable to the boarding of a spider that spins an egg sac immediately. There is thus a continuing emergence of adult *M. uhleri* throughout the summer, a given lineage producing one, two, or possibly, three generations every twelve months. The seasonal presence of adults in black-light traps in operation throughout the warm months of two consecutive years supports our theory of the emergence pattern of adult *M. uhleri*. Future studies will undoubtedly show that this insect plays a much greater role in the forest ecosystem than has been previously supposed.

Addendum

Several relevant papers dealing with the developmental and behavioral ecology of the Mantispidae have appeared during the three-year interim from the date of acceptance to the date of publication of this monograph. These papers are: Gilbert and Rayor (1983); Killebrew (1982); MacLeod and Redborg (1982); Opler (1981); Redborg (1982a, 1982b, 1983); Redborg and Mac-Leod (1983a, 1983b).

Appendix I. *Mantispa uhleri* Banks–Spider Associations in the Southern Illinois University at Carbondale Collection

Spider families, genera, and species	Number of spiders examined per family	Number of spiders per species Immature	♂	♀	Number of mantispid larvae removed	Months of collection
Antrodiaetidae	68					
Antrodiaetus unicolor (Hentz)		26	10	30		IV-VI,IX-XI
Atypoides hadros Coyle		2	0	0		X,XI
Ctenizidae	4					
Ummidia sp.		4	0	0		VI,VII,X
Amaurobiidae	12					
Amaurobius bennetti (Black-wall)		0	1	0		IV
Titanoeca americana Emerton		2	0	0		VI
Titanoeca brunnea Emerton		8	1	0		V-VII
Dictynidae	218					
Dictyna sp.		1	0	0		IV
Dictyna bellans Chamberlin		2	2	0		IV,VI-VIII
Dictyna bicornis Emerton		0	1	0		VI
Dictyna cruciata Emerton		9	25	0		V-VII
Dictyna foliacea (Hentz)		10	32	2		V-VII,IX
Dictyna formidolosa Gertsch and Ivie		2	0	0		V,VI
Dictyna hentzi Kaston		1	1	0		IV
Dictyna sublata (Hentz)		30	33	0		IV-VII,X
Dictyna volucripes Keyserling		11	40	10		IV-VIII,X
Lathys immaculata Chamberlin and Ivie		0	6	0		II
Oecobiidae	2					
Oecobius cellariorum (Duges)		1	1	0		V,VII
Uloboridae	10					
Uloborus glomosus (Walckenaer)		2	6	2		V,VII,VIII

Appendix I (cont.)

Spider families, genera, and species	Number of spiders examined per family	Number of spiders per species Immature	Number of spiders per species ♂	Number of spiders per species ♀	Number of mantispid larvae removed	Months of collection
Dysderidae	7					
Ariadna bicolor (Hentz)		0	2	3		IV,V,VIII,XI
Dysdera crocata C. L. Koch		0	0	2		IV
Scytodidae	48					
Loxosceles reclusa Gertsch and Mulaik		12	7	25		IV-X
Scytodes thoracica (Latreille)		0	3	1		IV,VI,VII,X
Pholcidae	22					
Pholcus phalangioides (Fuesslin)		5	9	6		II,V,VII,VIII,
Spermophora meridionalis (Hentz)		1	1	0		II,X
Theridiidae	810					
Achaearanea porteri (Banks)		5	3	7		V,VI,VIII,XI
Achaearanea globosa (Hentz)		3	2	2		V,VII,VIII
Achaearanea tepidariorum (C. L. Koch)		78	195	169		I,IV-XI
Argyrodes cancellatus (Hentz)		2	1	1		V-VII
Argyrodes elevatus Taczanomski		2	1	7		IV,V,IX
Argyrodes fictilium (Hentz)		0	0	1		X
Argyrodes trigonum (Hentz)		2	4	0		V,VI
Crustulina altera Gertsch and Archer		3	5	0		I,III,V,IX,XI
Ctenium frontatus (Banks)		1	3	0		I,VII,VIII
Dipoena nigra (Emerton)		2	6	0		V-VIII
Enoplognatha sp.		0	0	1		X
Euryopis funebris (Hentz)		5	0	2		IV,V,VIII
Euryopis quinquemaculata Banks		1	0	0		V
Latrodectus mactans (Fabricius)		2	15	0		V-X

Appendix I (cont.)

Spider families, genera, and species	Number of spiders examined per family	Number of spiders per species Immature	♂	♀	Number of mantispid larvae removed	Months of collection
Theridiidae (cont.)						
Latrodectus variolus Walckenaer	2	2	2			V
Nesticus pallidus Emerton	0	1	0			VI
Pholcomma hirsuta Emerton	2	1	2			III,V,XI
Phoroncidia americana (Emerton)	0	1	0			V
Spintharus flavidus Hentz	3	20	6			VII-IX
Steatoda borealis (Hentz)	4	12	9			IV,V,VIII,IX
Steatoda triangulosa (Walckenaer)	10	34	10			I,II,IV,V,VIII-XII
Stemmops ornatus (Bryant)	4	1	0			VI,VIII
Theridion alabamense Gertsch and Archer	2	1	0			IV-VI
Theridion albidum Banks	1	2	0			VI-VIII
Theridion antonii Keyserling	0	1	0			X
Theridion berkeleyi Emerton	0	1	0			V
Theridion differens Emerton	23	25	3			IV-X
Theridion flavonotatum Becker	1	2	0			V,VI
Theridion frondeum Hentz	4	15	2			VI,VII
Theridion glaucescens Becker	3	7	0			IV,VI,X,XI
Theridion lyricum Walckenaer	5	7	0			V,VIII-X
Theridion murarium Emerton	4	2	0			VI,VII,XII
Theridion pictipes Keyserling	5	21	1			V,VIII
Theridula opulenta (Walckenaer)	3	5	1			V,VII
Thymoites pallida (Emerton)	1	0	0			IX
Thymoites unimaculata (Emerton)	3	2	2			IV-VI

Appendix I (cont.)

Spider families, genera, and species	Number of spiders examined per family	Number of spiders per species Immature	σ	φ	Number of mantispid larvae removed	Months of collection
Linyphiidae	515					
Bathyphantes pallida (Banks)		0	3	1		V,XI
Centromerus cornupalpis (O.P.-Cambridge)		7	5	0		IV,X,XI
Centromerus latidens (Emerton)		3	8	1		I,III,X,XI
Ceraticelus creolus Chamberlin		1	3	0		IV,VI
Ceraticelus fissiceps (O.P.-Cambridge)		10	31	0		IV-IX
Ceraticelus laetabilis (O.P.-Cambridge)		0	1	0		III
Ceraticelus micropalpis (Emerton)		0	2	0		VIII,IX
Ceraticelus minutus (Emerton)		0	3	0		VI,X,XI
Ceratinella brunnea Emerton		1	0	0		IV
Ceratinopsidis formosa (Banks)		2	8	0		VIII,IX,X
Ceratinopsis sp.		0	1	0		V
Ceratinopsis laticeps Emerton		0	1	0		V
Ceratinopsis purpurescens (Keyserling)		4	10	8		IV-VII,X
Ceratinopsis tarsalis Emerton		1	0	0		VII
Cornicularia sp.		1	3	0		III,IV,XI
Cornicularia brevicornis Emerton		1	0	0		IX
Eperigone sp.		1	0	0		X
Eperigone banksi Ivie and Barrows		1	1	0		X
Eperigone maculata (Banks)		4	3	0		IV,V,VIII,XI
Eperigone tridentata (Emerton)		2	3	0		I,IV,VI,X

Appendix I (cont.)

Spider families, genera, and species	Number of spiders examined per family	Number of spiders per species		Number of mantispid larvae removed	Months of collection
		Im- mature	♂ ♀		
Linyphiidae (cont.)					
Eridantes erigonoides (Emerton)	0	4	0		IV-VI
Erigone autumnalis Emerton	15	5	0		III,V-VII,XI
Gonatium rubens (Blackwall)	0	1	0		XI
Grammonota inornata Emerton	2	3	0		VI,VII,X
Grammonota vittata Barrows	1	0	0		X
Islandiana flaveola (Banks)	2	0	0		VI,XII
Islandiana longisetosa (Emerton)	1	0	0		VII
Floricomus spp.	1	3	0		V,X,XI
Floricomus rostratus (Emerton)	1	0	0		VI
Florinda coccinea (Hentz)	4	14	3		IV,V,VII,X
Frontinella pyramitela (Walckenaer)	14	55	21		IV-VIII,X,XI
Linyphia spp.	1	1	0		VII,VIII
Linyphia maculata Emerton	1	1	0		VI,VIII
Linyphia marginata C.L. Koch	40	83	23		IV-X
Linyphia waldea Chamberlin and Ivie	1	0	0		V
Meioneta spp.	13	13	1		IV-X
Meioneta angulata (Emerton)	0	1	0		IX
Meioneta micaria (Emerton)	1	1	0		VIII,X
Meioneta unimaculata (Banks)	0	1	0		IX
Origanates rostratus (Banks)	3	14	0		I,III,V,IX-XI
Paracornicularia bicapillata Crosby and Bishop	1	2	0		XI
Phanetta subterranea Emerton	1	0	1		XII
Pityohyphantes costatus (Hentz)	0	2	7		IV,IX,X

Appendix I (cont.)

Spider families, genera, and species	Number of spiders examined per family	Number of spiders per species Immature	Number of spiders per species ♂	Number of spiders per species ♀	Number of mantispid larvae removed	Months of collection
Linyphiidae (cont.)						
Tapinopa bilineata Banks		0	0	1		VIII
Tennesseellum formicum (Emerton)		2	1	0		V-VII
Walckenaera vigilax (Blackwall)		1	0	0		VIII
unidentified		0	12	0		I,IV-VII,X,XI
Araneidae	1,141					
Acacesia hamata (Hentz)		6	13	2		VII,VIII,IX
Alpaida calix (Walckenaer)		0	1	0		V
Acanthepeira stellata (Marx)		9	5	6		IV-VI,IX,X
Araneus gigas (Leach)		0	1	0		VI
Araneus bonsallae (McCook)		1	2	1		VI
Araneus juniperi (Emerton)		1	0	0		VII
Araneus marmoreus Clerck		1	25	0		IX,X
Araneus niveus (Hentz)		0	1	0		IX
Araneus pratensis (Emerton)		10	9	5		IV-VI,VIII,X
Araneus thaddeus (Hentz)		0	2	0		X,XI
Araniella displicata (Hentz)		0	1	0		V
Argiope aurantia Lucas		5	8	1		VII-X,XII
Argiope trifasciata (Forskal)		3	12	0		IX,X
Cyclosa conica (Pallas)		1	0	0		IV
Cyclosa turbinata (Walckenaer)		1	2	0		V,VII,VIII
Gea heptagon (Hentz)		2	2	2		IV,V,VII,X
Hypsosinga funebris (Keyserling)		0	1	0		VI
Hypsosinga rubens (Hentz)		3	14	2		IV-VI,X
Hypsosinga pygmaea (Sundevall)		0	0	12		VII,VIII,X
Leucauge venusta (Walckenaer)		10	36	6		V-IX,XI
Mangora gibberosa (Hentz)		23	15	24		VI-IX
Mangora maculata (Keyserling)		13	35	2		VI-IX

Appendix I (cont.)

Spider families, genera, and species	Number of spiders examined per family	Number of spiders per species		Number of mantispid larvae removed	Months of collection
		Immature	♂ ♀		
Araneidae (cont.)					
Mungora placida (Hentz)	8	44	5		IV-VIII
Mastophora phrynosoma Gertch	0	1	0		VIII
Meta menardii (Latreille)	5	23	33		I,X,XI
Metepeira labyrinthea (Hentz)	6	15	1		VII,IX
Micrathena gracilis (Walckenaer)	6	82	3		III,VI-X
Micrathena mitrata (Hentz)	13	37	21		VII-X
Micrathena sagittata (Walckenaer)	1	11	1		VII-X
Mimognatha foxi (McCook)	3	3	0		VI,VII,X
Neoscona arabesca (Walckenaer)	8	36	5		IV,VI-XI
Neoscona domiciliorum (Hentz)	0	3	0		IX,X
Neoscona hentzii (Keyserling)	9	46	0		VII-XI
Nuctenea cornuta (Clerck)	19	41	3		IV-XI
Pachygnatha brevis Keyserling	0	1	0		?
Singa keyserlingi McCook	0	2	0		V,X
Tetragnatha elongata Walckenaer	25	28	14		VI-XI
Tetragnatha laboriosa Hentz	23	39	27		IV-X
Tetragnatha pallescens F.O.P.-Cambridge	0	6	0		VI,X
Tetragnatha seneca Seeley	0	11	27		VII,VIII,X
Tetragnatha straminea Emerton	5	9	21		V,VIII
Tetragnatha versicolor Walckenaer	1	3	0		VI-VIII,X
Verrucosa arenata (Walckenaer)	3	49	5		IV,VI-X

Appendix I (cont.)

Spider families, genera, and species	Number of spiders examined per family	Number of spiders per species Immature	of ♂	♀	Number of mantispid larvae removed	Months of collection
Araneidae (cont.)						
Wixia ectypa (Walckenaer)		3	1	1		VIII-X
unidentified		0	2	6		VII,X,XII
Symphytognathidae	2					
Maymena ambita (Barrows)		0	1	0		V
Mysmena guttata (Banks)		0	1	0		V
Mimetidae	18					
Ero sp.		0	1	0		IV
Ero furcata (Villers)		2	0	0		V,X
Mimetus epeiroides Emerton		0	0	1		VIII
Mimetus puritanus Chamberlin		1	6	0		VII,VIII
unidentified		6	1	0		VI-VIII
Agelendiae	248					
Agelenopsis emertoni Chamberlin and Ivie		16	37	46		VI-XI
Agelenopsis kastoni Chamberlin and Ivie		5	12	2		IV-VI,X
Agelenopsis naevia (Walckenaer)		2	10	0		VI,VIII-X
Agelenopsis pennsylvanica (C. L. Koch)		9	19	1		VI,VII,X
Cicurina arcuata Keyserling		10	15	0		I,III,IV-VIII,XI
Cicurina brevis (Emerton)		4	4	0		IV,V,IX-XI
Cicurina ludoviciana Simon		2	4	0		IV,VI,X,XI
Cicurina pallida Keyserling		0	2	0		IV,X
Coras lamellosus (Keyserling)		19	21	0		III-VII,IX-XI
Coras taugynus Chamberlin		0	1	0		V
Tegenaria domestica (Clerck)		3	2	2		IV,VI,VIII,X
Hahniidae	21					
Hahnia cinerea Emerton		1	1	0		X

Appendix I (cont.)

Spider families, genera, and species	Number of spiders examined per family	Number of spiders per species		Number of mantispid larvae removed	Months of collection
		Immature	♂	♀	

Spider families, genera, and species	Number of spiders examined per family	Immature	♂	♀	Number of mantispid larvae removed	Months of collection
Hahniidae (cont.)						
Hahnia flaviceps Emerton		0	7	1		III,IV,X
Neoantistea agilis (Keyserling)		8	3	0		III,V,VI,IX-XI
Pisauridae	76					
Dolomedes sp.		0	1	1		V,VI
Dolomedes albineus Hentz		0	1	1		IV,VI
Dolomedes scriptus Hentz		1	1	0		IV,VIII
Dolomedes tenebrosus Hentz		1	4	0	1	IV-VII,IX
Dolomedes triton (Walckenaer)		4	4	4		IV,V,VII,IX
Dolomedes vittatus Walckenaer		0	1	3		V,X
Pisaurina sp.		0	2	1		II
Pisaurina brevipes (Emerton)		12	13	2		II-V
Pisaurina dubia (Hentz)		5	1	0		II,IV,V
Pisaurina mira (Walckenaer)		4	6	3		IV-VI,IX
Lycosidae	614					
Arctosa funerea (Hentz)		2	2	0		VI,IX,X
Geolycosa missouriensis (Banks)		0	1	0		II
Lycosa sp.		0	0	1		IV
Lycosa antelucana Montgomery		0	1	0		IV
Lycosa aspersa Hentz		0	2	0		V
Lycosa baltimoriana (Keyserling)		0	1	0		VI
Lycosa carolinensis Walckenaer		7	4	0		V,IX-XI
Lycosa frondicola Emerton		0	3	1		IV,V
Lycosa georgicola Walckenaer		1	2	0		VIII,X
Lycosa gulosa Walckenaer		3	6	0		III,IV,IX-XI
Lycosa helluo Walckenaer		3	9	5		V-VIII,X,XI
Lycosa hentzi Banks		0	0	2		IV,IX
Lycosa pulchra (Keyserling)		4	2	0		II,X,XI

Appendix I (cont.)

Spider families, genera, and species	Number of spiders examined per family	Number of spiders per species Immature		Number of mantispid larvae removed	Months of collection
		Im- mature ♂	♀		
Lycosidae (cont.)					
Lycosa punctulata Hentz		6	13 10		IV,V,IX,X
Lycosa rabida Walckenaer		9	12 1		VI-X
Pardosa milvina (Hentz)		29	66 13		IV-XI
Pirata spp.		1	0 1		V,X
Pirata alachua Gertsch and Wallace		31	29 3		V-IX
Pirata insularis Emerton		0	1 0		VI
Pirata maculatus Emerton		3	3 0		VI,VIII,IX,X
Pirata spiniger (Simon)		1	0 2		IX
Schizocosa sp.		0	1 0		VI
Schizocosa avida (Walckenaer)		5	9 0		IV-VII
Schizocosa bilineata (Emerton)		23	3 1		V,VI
Schizocosa ocreata (Hentz)		85	83 5		IV-X
Schizocosa saltatrix (Hentz)		32	13 0		IV-VII
Trochosa avara Keyserling		15	19 0		III-VI,X,XI
unidentified		5	9 10		IV-VIII,X,XI
Oxyopidae	289				
Oxyopes sp.		2	1 0		VII
Oxyopes aglossus Chamberlin		5	3 0		V-VIII
Oxyopes scalaris Hentz		0	0 1		IX
Oxyopes salticus Hentz		68	69 140		V-VIII,X
Gnaphosidae	108				
Cesonia bilineata (Hentz)		3	0 0		V,VI
Drassylus aprilinus (Banks)		6	2 2		II,III,V,VI,XI
Drassylus covensis Exline		12	2 0		V,VI
Drassylus creolus Chamberlin and Gertsch		1	1 0		V
Drassylus depressus (Emerton)		1	0 0		V
Drassylus fallens Chamberlin		1	1 0		VI

Appendix I (cont.)

Spider families, genera, and species	Number of spiders examined per family	Number of spiders per species		Number of mantispid larvae removed	Months of collection
		Immature	♂ ♀		
Gnaphosidae (cont.)					
Drassylus frigidus (Banks)		1	0 0		II
Drassylus virginianus Chamberlin		9	2 0		V,VI,IX
Gnaphosa fontinalis Keyserling		1	2 0		V-VII
Gnaphosa sericata (L. Koch)		4	2 0		VI,VII
Haplodrassus bicornis (Emerton)		3	1 0		V,VI
Haplodrassus signifer (C. L. Koch)		2	3 0		IV VI
Herpyllus ecclesiasticus Hentz		8	8 0		III-VIII,X
Litopyllus rupicolens Chamberlin		0	1 0		VIII
Micaria laticeps Emerton		3	0 0		IV,V
Rachodrassus exlinae Platnick and Shaalab		1	1 1		III,VI
Sergiolus sp.		0	0 1		V
Sergiolus decoratus Kaston		0	1 0		IX
Sergiolus famulus Chamberlin		1	0 0		VI
Sergiolus variegatus (Hentz)		0	0 1		X
Zelotes duplex Chamberlin		4	2 0		IV-VI
Zelotes hentzi Barrows		5	1 0		III-V,X
unidentified		2	4 1		V,VII
Clubionidae	350				
Castianeira amoena (C. L. Koch)		0	1 0		VII
Castianeira cingulata (C. L. Koch)		4	11 0		III,V-IX
Castianeira crocata (Hentz)		1	4 0		V-VII
Castianeira descripta (Hentz)		0	0 1		IV

Appendix I (cont.)

Spider families, genera, and species	Number of spiders examined per family	Number of spiders per species Immature	Number of spiders per species ♂	Number of spiders per species ♀	Number of mantispid larvae removed	Months of collection
Clubionidae (cont.)						
Castianeira longipalpus (Hentz)		2	5	2		V,VI,IX-XI
Castianeira trilineata (Hentz)		1	0	0		IV
Castianeira variata Gertsch		1	1	0		VII,VIII
Chiracanthium inclusum (Hentz)		1	0	0		VII
Chiracanthium mildei L. Koch		7	6	1	1	IV,V,VII,IX-X
Clubiona abbotii L. Koch		4	12	8		V-VIII,X
Clubiona catawba Gertsch		4	3	2		IV-VI
Clubiona excepta L. Koch		15	9	0		V-XI
Clubiona kastoni Gertsch		0	4	0		V,VI
Clubiona maritima L. Koch		4	3	1		VI,VIII
Clubiona obesa Hentz		3	4	0	1	IV-VI,X
Clubiona pygmaea Banks		1	0	0		VI
Phrurotimpus alarius (Hentz)		51	28	65		II-IX,XI,XII
Phrurotimpus illudens Gertsch		9	0	0		V-VII
Scotinella sp.		0	5	0		V,IX,X
Scotinella britcheri (Petrunkevitch)		2	1	0		IV,V,X
Scotinella formica (Banks)		1	2	0		II,VIII
Scotinella goodnighti (Muma)		18	27	0		II-XI
Scotinella redempta (Gertsch)		3	6	0		III,V,VIII,X
Strotarchus piscatorius (Hentz)		1	1	0		V,VI
Trachelas deceptus (Banks)		1	2	0		IV,V
unidentified		0	1	0		VII
Anyphaenidae	118					
Anyphaena sp.		0	0	5	1	X

Appendix I (cont.)

Spider families, genera, and species	Number of spiders examined per family	Number of spiders per species		Number of mantispid larvae removed	Months of collection
		Im-mature	♂ ♀		
Anyphaenidae (cont.)					
Anyphaena celer (Hentz)		1	1 0		X
Anyphaena fraterna (Banks)		5	3 0		IV-VII,X
Anyphaena maculata (Banks)		1	1 0		X
Anyphaena pectorosa L. Koch		6	6 0		VI-VIII
Aysha sp.		0	0 27		IV
Aysha gracilis (Hentz)		6	6 1		IV,V,VII,VIII
Wulfila alba (Hentz)		1	4 0		VI,VIII
Wulfila saltabunda (Hentz)		3	17 24		V-VII
Ctenidae	2				
Anahita animosa (Walckenaer)		0	0 1		IV
Zora pumila (Hentz)		0	0 1		III
Thomisidae	406				
Coriarachne floridana Banks		0	1 0		IV
Coriarachne versicolor Keyserling		1	1 0		IV,X
Ebo sp.		0	0 1		XII
Ebo latithorax Keyserling		1	0 1		V,XI
Misumenoides aleatorius (Hentz)		9	16 10		IV,VI-X
Misumenops asperatus (Hentz)		14	13 1		IV-VII,IX
Misumenops celer (Hentz)		1	7 0	1	IV-VI,IX,X
Misumenops oblongus (Keyserling)		11	20 1		V-VIII,XI
Oxyptila monroensis Keyserling		0	4 5		I,IV-VI
Philodromus bimuricatus Dondale and Redner		1	2 0		IV,VI
Philodromus infuscatus Keyserling		0	2 0		X,XI

Appendix I (cont.)

Spider families, genera, and species	Number of spiders examined per family	Number of spiders per species Immature	♂	♀	Number of mantispid larvae removed	Months of collection
Thomisidae (cont.)						
Philodromus keyserlingi Marx	5	7	0			VI,VII
Philodromus laticeps Keyserling	1	0	0			XII
Philodromus marxii Keyserling	5	3	0			IV-VII
Philodromus minutus Banks	3	2	0			V,VI
Philodromus montanus Bryant	1	0	0			IV
Philodromus placidus Banks	2	0	0			VI
Philodromus pratariae (Scheffer)	15	8	7			VIII,IX
Philodromus rufus Walckenaer	4	1	0			IV-VI
Philodromus vulgaris (Hentz)	5	5	1	1		II,IV-VII,X
Synema sp.	1	1	0			V,VI
Synema parvula (Hentz)	40	32	42			IV-X
Tibellus duttoni (Hentz)	1	2	1	1		II,VI
Tibellus oblongus (Walckenaer)	0	1	0			II
Thanatus rubicellus Mello-Leitao	1	0	0			X
Tmarus angulatus (Walckenaer)	0	5	1			VI,VIII
Xysticus auctificus Keyserling	0	7	0			V-VII,XI
Xysticus discursans Keyserling	1	0	0			IV
Xysticus elegans Keyserling	4	0	0			IV,VI
Xysticus fraternus Banks	16	3	0			V,VI
Xysticus funestus Keyserling	13	14	1			IV-VI,VIII-XII
Xysticus ferox (Hentz)	9	8	1			V-VII,X

Appendix I (cont.)

Spider families, genera, and species	Number of spiders examined per family	Number of spiders per species		Number of mantispid larvae removed	Months of collection
		Im-mature	♂ ♀		
Thomisidae (cont.)					
Xysticus texanus Banks		1	1 0		VII
Xysticus triguttatus Keyserling		0	1 0		VI
Salticidae	654				
Agassa cyanea (Hentz)		4	6 0		V-VII
Ballus youngii Peckham		2	0 0		IV
Evarcha hoyi (Peckham)		0	3 0		V,VIII
Gertschia noxiosa (Hentz)		0	2 0		V
Habrocestum pulex (Hentz)		7	7 2		V,VI,XI
Habronattus sp.		0	0 1		VIII
Habronattus agilis (Banks)		1	0 0		IV
Habronattus decorus (Blackwall)		0	1 0		VI
Hentzia mitrata (Hentz)		3	10 19	2	IV-VII,X,XI
Hentzia palmarum (Hentz)		5	3 0		V,VI
Icius elegans (Hentz)		1	7 0		VI,VII
Icius hartii Emerton		1	0 0		VI
Icius similis Banks		0	1 0		VI
Maevia inclemens (Walckenaer)		9	5 0		I,V-VII
Marpissa formosa (Banks)		5	4 2		VI,VII
Marpissa lineata (C. L. Koch)		1	1 0		IV,V
Marpissa pikei (Peckham)		12	11 0		IV-VI,VIII
Metacyrba undata (De Geer)		12	16 1		IV-VI,IX,XI
Metaphidippus spp.		17	13 0		III,IV-VI,X
Metaphidippus canadensis (Banks)		0	2 0		V,X
Metaphidippus galathea (Walckenaer)		43	31 14		IV-VII
Metaphidippus protervus (Walckenaer)		94	22 40	3	IV-VII,IX,X
Paraphidippus aurantius (Lucas)		12	6 1	1	V-VII,X

Appendix I (cont.)

Spider families, genera, and species	Number of spiders examined per family	Number of spiders per species Immature		Number of mantispid larvae removed	Months of collection
		Immature ♂	♀		
Thomisidae (cont.)					
Paraphidippus marginatus (Walckenaer)	12	9	0		IV,V,VII,VIII
Phidippus sp.	0	0	2		IX
Phidippus audax (Hentz)	20	11	2	2	I-VI
Phidippus clarus Keyserling	11	8	5		VI-VIII
Phidippus mystaceus (Hentz)	0	1	1		VIII,IX
Phidippus princeps (Peckham)	4	7	0		IV-VI
Phidippus putnamii (Peckham)	1	0	0	1	VIII
Phlegra fasciata (Hahn)	2	0	0		V
Salticus scenicus (Clerck)	1	3	2		III-VI
Sassacus papenhoei Peckham	0	1	1		VI
Sitticus cursor Barrows	1	0	0		V
Sitticus fasciger (Simon)	2	3	0		I,III,IV,XI,XII
Synemosyna formica Hentz	0	0	9		IX,X
Thiodina sp.	5	0	0		IV,VI
Thiodina puerpera (Hentz)	5	4	0		IV-VII
Thiodina sylvana (Hentz)	14	13	5		IV-X
Zygoballus bettini Peckham	10	10	0		IV-VII,X
Zygoballus nervosus (Peckham)	0	3	0		VII-IX
Zygoballus sexpunctatus (Hentz)	1	5	0		IV-VI,X
Totals	5,761	5,761		16	

Appendix II. Spider Species Utilized by First Instar *Mantispa uhleri* Larvae†

Agelenidae
Agelenopsis kastoni Chamberlin and Ivie
female: left book lung; IV-15-76

Anyphaenidae
Anyphaena sp.
immature male: left side of pedicel; Union County*; X-7-67
Anyphaena fraterna (Banks)
male: dorsal pedicel in left pit; IV-16-76
male: left side of pedicel; IV-17-76
Anyphaena pectorosa L. Koch
female: pedicel; IX-4-74

Clubionidae
Chiracanthium mildei L. Koch
immature female: right book lung; Jackson County*; IV-21-68
Clubiona obesa Hentz
male: dorsal pedicel; IV-16-76
female: in egg sac; Williamson County*; IV-18-71

Gnaphosidae
Herpyllus ecclesiasticus Hentz
male: right side of pedicel; II-8-76

Lycosidae
Lycosa pulchra (Keyserling)
immature: pedicel; VIII-6-74
immature: right side of pedicel; VIII-7-74
immature: ventral pedicel; VIII-7-74
immature female: book lung; VIII-12-74
male: left side of pedicel; I-8-75

†Each line entry beneath a species name gives data (sex of spider: position of larva; collection locale; date of collection) for an individual of that species from which at least one larva of *M. uhleri* was removed. Multiple larvae from a single spider are recorded by a number. An asterisk following the collection locale signifies that that spider is from the SIUC Collection; if no locale is indicated, the collection was made at DSAC (see also p. 76).

117

Lycosa punctulata Hentz
  immature female: right book lung; IX-4-74
  male: right side of pedicel; IX-10-74
Lycosa rabida Walckenaer
  male: dorsal pedicel in left pit; VIII-7-74
Schizocosa ocreata (Hentz)
  female: left edge of carapace; VIII-6-74
  female: right side of pedicel; VIII-7-74

Pisauridae
  Dolomedes tenebrosus Hentz
  male: left book lung; Pope County*; V-19-71
Pisaurina mira (Walckenaer)
  immature male: dorsal pedicel in left pit; IX-29-74
  immature male: ventral pedicel; I-28-75
  immature female: 2 larvae—right side of pedicel, right book lung;
    IX-10-74

Salticidae
  Evarcha hoyi (Peckham)
  female: right book lung; X-16-74
Hentzia mitrata (Hentz)
  immature male: dorsal pedicel; Jackson County*; X-13-66
  immature female: dorsal pedicel; IV-18-76
  female: ventral pedicel; Hamilton County*; VII-20-72
Maevia inclemens (Walckenaer)
  immature male: pedicel; IX-3-74
Metacyrba sp.
  immature female: dorsal pedicel; XI-13-74
Metacyrba undata (De Geer)
  immature: left side of pedicel; I-23-75
  immature: left side of pedicel; I-23-75
  immature: right side of pedicel; II-4-75
  immature: ventral pedicel; II-4-75
  immature: right side of pedicel; II-4-75
  immature: dorsal pedicel; II-14-75
  immature: ventral pedicel; II-14-75
  immature: ventral pedicel; II-14-75
  immature: left side of pedicel; II-14-75
  immature: right side of pedicel; II-8-76
  immature: ventral pedicel; II-8-76
  immature: ventral pedicel; II-26-76

Salticidae (cont.)
  Metacyrba undata (De Geer) (cont.)
    immature: ventral pedicel; III-4-76
    immature: left side of pedicel; III-15-76
    immature: right side of pedicel; III-15-76
    immature: right side of pedicel; III-15-76
    immature: left side of pedicel; III-16-76
    immature: dorsal pedicel in left pit; III-16-76
    immature: left side of pedicel; III-16-76
    immature: right side of pedicel; III-25-76
    immature: right side of pedicel; III-25-76
    immature male: pedicel; Alexander County; I-15-74
    immature male: ventral pedicel; II-4-75
    male: pedicel; Alexander County; I-15-74
    male: pedicel; Hardin County; I-16-74
    male: pedicel; Hardin County; I-16-74
    male: left side of pedicel; I-7-75
    male: right side of pedicel; I-7-75
    male: ventral pedicel; I-23-75
    male: ventral pedicel; I-31-75
    male: pedicel; II-2-75
    male: right side of pedicel; II-3-75
    male: dorsal pedicel; II-14-75
    male: right side of pedicel; II-14-75
    male: left side of pedicel; II-14-75
    male: ventral pedicel; II-14-75
    male: dorsal pedicel; III-15-76
    male: left side of pedicel; III-15-76
    male: right side of pedicel; III-16-76
    male: ventral pedicel; III-16-76
    female: pedicel; Alexander County; I-15-74
    female: pedicel; Alexander County; I-15-74
    female: ventral pedicel; I-7-75
    female: dorsal pedicel in right pit; II-14-75
    female: dorsal pedicel in left pit; II-14-75
    female: ventral pedicel; III-15-76
    female: spinnerets; III-16-76
  Metaphidippus galathea (Walckenaer)
    female: pedicel; V-12-75
  Metaphidippus protervus (Walckenaer)
    male: ventral pedicel; Jackson County*; X-15-66

male: ventral pedicel; Pope County*; V-19-71
male: left side of pedicel; Pope County*; V-19-71
Paraphidippus aurantius (Lucas)
    male: ventral pedicel; Jackson County*; VI-18-71
Pellenes sp.
    female: ventral pedicel; IX-17-74
Phidippus audax (Hentz)
    immature: dorsal pedicel; IX-12-74
    immature: between leg bases and carapace; IX-17-74
    immature: right book lung; IX-27-74
    immature: left side of pedicel; X-12-74
    immature: 2 larvae—ventral pedicel, left book lung; XI-9-74
    immature male: right side of pedicel; IX-17-74
    immature male: right side of pedicel; IX-17-74
    immature male: left side of pedicel; IX-17-74
    immature male: dorsal pedicel in left pit; X-10-74
    immature male: right side of pedicel; X-10-74
    immature male: dorsal pedicel in left pit; X-11-74
    immature male: left side of pedicel; X-11-74
    immature female: 3 larvae—ventral pedicel, dorsal pedicel in left
        pit, between leg bases and carapace; IX-7-74
    immature female: 3 larvae—between sternum and leg bases, dorsal
        pedicel in right pit, left book lung; IX-19-74
    immature female: ventral pedicel; IX-20-74
    immature female: left book lung; IX-20-74
    immature female: dorsal pedicel; X-6-74
    immature female: dorsal pedicel in left pit; X-10-74
    immature female: ventral pedicel; I-10-75
    male: ventral pedicel; Pope County*; V-19-71
    male: right book lung; VI-16-75
    female: dorsal pedicel in right pit; Pope County*; V-19-71
Phidippus clarus Keyserling
    immature: left book lung; IX-19-74
    immature: pedicel; IX-20-74
Phidippus princeps (Peckham)
    immature male: right book lung; IX-20-74
    immature female: right book lung; IX-20-74
Phidippus putnamii (Peckham)
    male: on abdomen just above pedicel; Jackson County*; VIII-30-no
        year

Phidippus whitmanii Peckham
   immature female: left book lung; IV-2-76
   immature female: dorsal pedicel; IV-4-76

Thomisidae
   Misumenoides aleatorius (Hentz)
      female: 2 larvae—left side of pedicel, left book lung; IX-3-74
   Misumenops sp.
      immature: pedicel; IX-24-74
   Misumenops celer (Hentz)
      female: in egg sac; Pope County*; V-19-71
   Philodromus vulgaris (Hentz)
      female: in egg sac; Jackson County*; VI-5-71
      female: dorsal pedicel; III-16-76
   Tibellus duttoni (Hentz)
      immature female: on left abdomen just above pedicel; Jackson
         County*; no date
   Xysticus ferox (Hentz)
      female: right book lung; IV-15-76
   Xysticus funestus Keyserling
      male: 2 larvae—ventral pedicel, between leg bases; IX-10-74

Literature Cited

Batra, S. W. T. 1972. Notes on the behavior and ecology of the mantispid, *Climaciella brunnea occidentalis*. J. Kansas Entomol. Soc. 45: 334-40.

Beatty, J. A., and J. M. Nelson. 1979. Additions to the checklist of Illinois spiders. Great Lakes Entomol. 12: 49-56.

Birabén, M. 1960. *Mantispa* (Neuroptera) parásita en el cocón de *Metepeira* (Araneae). Neotropica 6: 61-64.

Bisset, J. L., and V. C. Moran. 1967. The life history and cocoon spinning behavior of a South African mantispid (Neuroptera: Mantispidae). J. Entomol. Soc. Southern Africa 30: 82-95.

Borror, D. J., and D. M. DeLong. 1971. An introduction to the study of insects. 3rd ed. Holt, Rinehart and Winston, New York.

Brauer, F. 1852. Verwandlungsgeschichte der *Mantispa pagana*. Archiv für Naturgeschichte 18: 1-2.

Brauer, F. 1855. Beiträge zur Kenntniß der Verwandlung der Neuropteren. Verh. Zool.-Bot. Ges. Wien 5: 479-84.

Brauer, F. 1869. Beschreibung der Verwandlungsgeschichte der *Mantispa styriaca* Poda and Betrachtungen über die sogenannte Hypermetamorphose Fabre's. Verh. Zool.-Bot. Ges. Wien 19: 831-40.

Bristowe, W. S. 1932. *Mantispa*, a spider parasite. Entomol. Mon. Mag. 68: 222-24.

Capocasale, R. 1971. Hallazgo de *Mantispa decorata* Erichson parasitando la ooteca di una *Lycosa poliostoma* (Koch) (Neuroptera, Mantispidae; Araneae, Lycosidae). Rev. Brasileira Biol. 31: 367-70.

Davidson, J. A. 1969. Rearing *Mantispa viridis* Walker in the laboratory (Neuroptera: Mantispidae). Entomol. News 80: 29-31.

Eltringham, H. 1932. On an extrusible glandular structure in the abdomen of *Mantispa styriaca* Poda (Neuroptera). Trans. Entomol. Soc. London 80: 103-105.

George, L. D., and N. L. George. 1975. Notes on the three hundred and fifty-eighth meeting: A new record of mantispid reared from spider. Pan-Pacific Entomol. 51: 90.

Gilbert, C., and L. S. Rayor. 1983. First record of mantisfly (Neuroptera: Mantispidae) parasitizing a spitting spider (Scytodidae). J. Kansas Entomol. Soc. 56: 578-80.

Handschin, E. 1960. Beiträge zu einer Revision der Mantispiden (Neuroptera). II. Teil. Mantispiden des "Musée Royal du Congo Belge" Tervuren. Rev. Zool. Bot. Africaines 62: 181-245.

Hoffman, C. H. 1936. Notes on *Climaciella brunnea* var. *occidentalis* Banks (Mantispidae—Neuroptera). Bull. Brooklyn Entomol. Soc. 31: 202-203.

Huber, I. 1958. Color as an index to the relative humidity of plaster of Paris culture jars. Proc. Entomol. Soc. Washington 60: 289-91.

Hungerford, H. B. 1936. The Mantispidae of the Douglas Lake, Michigan, region with some biological observations. Entomol. News 47: 69-72, 85-88.

Hungerford, H. B. 1939. A note on Mantispidae. Bull. Brooklyn Entomol. Soc. 34: 265.

Kaston, B. J. 1938. Mantispidae parasitic on spider egg sacs. J. New York Entomol. Soc. 46: 147-53.

Kaston, B. J. 1940. Another *Mantispa* reared. Bull. Brooklyn Entomol. Soc. 35: 21.

Kaston, B. J. 1948. Spiders of Connecticut. State of Connecticut, Hartford.

Killebrew, D. W. 1982. *Mantispa* in a *Peucetia* egg case. J. Arachnol. 10: 281-82.

Killington, F. J. 1936. A monograph of the British neuroptera. Vol. I. The Ray Society, London.

Kishida, K. 1929. On the oviposition of a clubionid spider, *Chiracanthium rubicundulum*. Lansania 1: 73-74 (In Japanese).

Kuroko, H. 1961. On the eggs and first-instar larvae of two species of Mantispidae. Esakia 3: 25-32.

Lucchese, E. 1955. Richerche sulla *Mantispa perla* Pallas (Neuroptera Planipennia—Fam. Mantispidae). Ann. Fac. Agrar. Univ. Stud. Perugia 11: 242-62.

Lucchese, E. 1956. Richerche sulla *Mantispa perla* Pallas (Neuroptera Planipennia—Fam. Mantispidae). Ann. Fac. Agrar. Univ. Stud. Perugia 12: 83-213.

McKeown, K. C., and V. H. Mincham. 1948. The biology of an Australian mantispid (*Mantispa vittata* Guerin). Australian Zool. 11: 207-24.

MacLeod, E. G., and K. E. Redborg. 1982. Larval platymantispine mantispids (Neuroptera: Planipennia): Possibly a subfamily of generalist predators. Neuroptera Intern. 2: 37-41.

124   *Mantispa uhleri* Banks

MacLeod, E. G., and P. Spiegler. 1961. Notes on the larval habitat and development peculiarities of *Nallachius americanus* (McLachean) (Neuroptera: Dilaridae). Proc. Entomol. Soc. Washington 63: 281-86.

Maerz, A., and M. R. Paul. 1930. A dictionary of color, 1st ed. McGraw-Hill, New York.

Main, H. 1931. A preliminary note on *Mantispa*. Proc. Entomol. Soc. London 6: 26.

Metcalf, R. L., and W. Luckmann. 1975. Introduction to insect pest management. John Wiley and Sons, New York.

Milliron, H. E. 1940. The emergence of a neotropical mantispid from a spider egg sac. Ann. Entomol. Soc. Am. 33: 357-60.

Nesbitt, H. H. 1945. A revision of the family Acaridae (Tyroglyphicae), order Acari, based on comparative morphological studies. Part I. Historical, morphological, and general taxonomic studies. Canadian J. Research Section D Zoological Sciences 23: 139-188.

New, T. R., and A. J. Haddow. 1973. Nocturnal flight activity of some African Mantispidae (Neuroptera). J. Entomol. Ser. A Gen. Entomol. 47: 161-68.

Opler, P. A. 1981. Polymorphic mimicry of polistine wasps by a neotropical neuropteran. Biotropica 13: 165-76.

Parfin, S. 1958. Notes on the bionomics of the Mantispidae (Neuroptera: Planipennia). Entomol. News 19: 203-207.

Peck, W. B., and W. H. Whitcomb. 1978. The phenology and populations of ground surface, cursorial spiders in a forest and a pasture in the south central United States. Symp. Zool. Soc. London 42: 131-38.

Poujade, G. A. 1898. Observation sur les moeurs de *Mantispa styriaca* Poda. Bull. Soc. Entomol. France 3: 347.

Redborg, K. E. 1982a. Mantispidae parasitic on spider egg sacs: An update of a pioneering paper by B. J. Kaston. J. Arachnol. 10: 92-93.

Redborg, K. E. 1982b. Interference by the mantispid *Mantispa uhleri* with the development of the spider *Lycosa rabida*. Ecol. Entomol. 7: 187-96.

Redborg, K. E. 1983. A mantispid larva can preserve its spider egg prey: Evidence for an aggressive allomone. Oecologia 58: 230-31.

Redborg, K. E., and E. G. MacLeod. 1983a. *Climaciella brunnea* (Say) (Neuroptera: Mantispidae): A mantispid that obligately boards spiders. J. Nat. Hist. 17: 63-73.

Redborg, K. E., and E. G. MacLeod. 1983b. Maintenance feeding of first instar mantispid larvae (Neuroptera, Mantispidae) on spider (Arachnida, Araneae) hemolymph. J. Arachnol. 11: 337-41.

Rogenhofer, A. 1862. Beitrag zur Kenntnis der Entwicklungsgeschichte von *Mantispa styriaca* Poda (*pagana* Fab.). Verh. Zool.-Bot. Ges. Wien. 12: 583-86.

Sheldon, J. K., and E. G. MacLeod. 1971. Studies on the biology of the Chrysopidae II. The feeding behavior of the adult of *Chrysopa carnea* (Neuroptera). Psyche J. Entomol. 78: 107-21.

Smith, R. C. 1922. The biology of the Chyrsopidae. Cornell Univ. Agr. Exp. Sta. Mem. 58: 1287-1372.

Smith, R. C. 1934. Notes on the Neuroptera and Mecoptera of Kansas, with keys for the identification of species. J. Kansas Entomol. Soc. 7: 120-45.

Stein, R. J. 1955. An insect masquerader. Nat. Hist. 11: 472-73.

Valerio, C. E. 1971. Parasitismo en heuvos de araña *Achaeuranea tepidariorum* (Koch) (Aranea: Theridiidae) en Cost Rica. Rev. Biol. Trop. 18: 99-106.

Viets, D. 1941. A biological note on the Mantispidae (Neuroptera) J. Kansas Entomol. Soc. 14: 70-71.

Withycombe, C. L. 1925. Some aspects of the biology and morphology of the Neuroptera. With special reference to the immature stages and their possible phylogenetic significance. Trans. Entomol. Soc. London 72: 303-411.

Index

A Note on the Authors

KURT E. REDBORG received his Ph.D. from the University of Illinois, at which he has subsequently been a Research Associate in the Department of Agronomy, a Visiting Specialist in the Department of Physiology and Biophysics, and now is Education Services Coordinator at the Thames Science Center in New London, Connecticut. Extensively interested in insects throughout his youth, he collected and became intrigued with the family Mantispidae while serving as a teaching assistant for a course in insect taxonomy. He is particularly interested in behavioral ecology and evolutionary theory, and plans to use the family Mantispidae as a tool in their study. This is his first book. It was awarded the 1981 Balduf Research Award by the Department of Entomology for excellence in entomological research and publication among graduate students.

ELLIS G. MacLEOD received his Ph.D. from Harvard University and is currently Associate Professor of Entomology/Genetics and Development at the University of Illinois. He is interested in all aspects of the evolution of insects. These include the mechanisms of speciation, as well as an analysis of the broad-scale features of the adaptive radiation of the insect orders as deduced from the fossil record and from comparative studies of living species. His research has dealt with numerous facets of the biology of the Neuroptera, including their reproductive cycles and an account of the taxonomy and zoogeography of the extinct Neuroptera of the Baltic amber.